Preface

Instrumental techniques of analysis have now moved from the confines of the chemistry laboratory to form an indispensable part of the analytical armoury of many workers involved in the biological sciences. It is now quite out of the question to consider a laboratory dealing with the analysis of biological materials that is not equipped with an extensive range of instrumentation. Recent years have also seen a dramatic improvement in the ease with which such instruments can be used, and the quality and quantity of the analytical data that they can produce. This is due in no small part to the ubiquitous use of microprocessors and computers for instrumental control. However, under these circumstances there is a real danger of the analyst adopting a 'black box' mentality and not treating the analytical data produced in accordance with the limitations that may be inherent in the method used. Such a problem can only be overcome if the operator is fully aware of both the theoretical and instrumental constraints relevant to the technique in question. As the complexity and sheer volume of material in undergraduate courses increases, there is a tendency to reduce the amount of fundamental material that is taught prior to embarking on the more applied aspects. This is nowhere more apparent than in the teaching of instrumental techniques of analysis. Clearly there is not sufficient time to study the molecular basis of all the techniques in depth, but unless the fundamental principles are understood, the analyst will never be able to develop a critical approach to an analytical problem, nor will he/she be able to evaluate the significance of the data produced. There are many excellent texts describing the various instrumental techniques, some of which are cited in this volume. However, they would appear to be more suitable for chemistry undergraduates rather than for students working in the applied biological sciences. The present volume attempts to provide such students with an adequate theoretical background of the techniques most relevant to the biological sciences and to show how the techniques may be applied to a wide range of analytical problems.

The techniques discussed may be simply divided into separative (liquid and gas chromatography, electrophoresis) and non-separative (various spectrometric techniques, flame techniques and electrochemical methods). In all of the chapters the theoretical basis has been discussed as far as is necessary to understand the technique in question. This naturally leads to a somewhat uneven treatment from chapter to chapter, for example the theoretical aspects of nuclear magnetic resonance are conceptually far more complex than those involved in chromatographic separations. No attempt has been made to cover all the possible analytical techniques which find application in the biological sciences, as this would have resulted either in a very superficial coverage of the techniques or in an excessively long book. The book has rather concentrated on those techniques considered to be the most important from a qualitative and quantitative analytical point of view.

It is hoped that the book will find a wide audience amongst those students studying a wide range of biologically based subjects, including biochemistry, medicine, nutrition, agricultural chemistry, pharmaceutical chemistry, environmental science and food science. It may well also prove useful to those scientists embarking on research in the biological sciences without a grounding in instrumental techniques, including those working in industry and other scientific establishments.

MHG
RM

Contents

Abbreviations and symbols

A	Absorbance
AAS	Atomic absorption spectrometry
AFS	Atomic fluorescence spectrometry
α	Degree of ionization (alpha)
ATP	Adenosine triphosphate
ATR	Attenuated total reflectance
AUFS	Absorbance units full scale
B, B_0, B_1	Applied magnetic field
β	Phase ratio (high-performance liquid chromatography); Bohr magneton (ESR) (beta)
BOD	Biochemical oxygen demand
c	Mass transfer term
c	Velocity of light in a vacuum
CIE	Commission Internationale de l'Eclairage
CM	Carboxymethyl
CT	Computed tomography
CW	Continuous wave
d	Separation of teeth in a diffraction grating
D_g	Diffusion coefficient in the gas phase
D_l	Diffusion coefficient in the liquid phase
δ	Chemical shift (delta)
DEAE	Diethylamino-ethyl
e	Electronic charge
E	Energy of atomic or molecular state; kinetic energy (mass spectrometry)
E_i	Energy of ith energy level
ΔE	Difference in energy between states
ECD	Electron capture detector (gas chromatography); electrochemical detector (electrochemistry)
ELDOR	Electron double resonance
ENDOR	Electron nuclear double resonance
EPR	Electron paramagnetic resonance
ε	Molar absorptivity (absorption coefficient) (epsilon)
ε^0	Solvent strength based on heat of adsorption
ESR	Electron spin resonance
eV	Electron volt
η	Viscosity (eta)
F	Force
[F]	Concentration of ground state fluorescent molecule
[F*]	Concentration of first excited singlet
FES	Flame emission spectrometry
FID	Flame ionization detector (gas chromatography); free induction decay (nuclear magnetic resonance spectroscopy)
FPD	Flame photometric detector
FTIR	Fourier transform infrared
g	Landé g-factor, spectroscopic splitting factor (electron spin resonance)
G	Free energy
γ	Magnetogyric ratio (gamma)

GC	Gas chromatography
GLC	Gas–liquid chromatography
GSC	Gas–solid chromatography
H	Height equivalent to a theoretical plate
ΔH	Enthalpy change
ΔH_x	Enthalpy of solvation of component x
h	Peak height (high-performance liquid chromatography, gas chromatography); Planck's constant
h'	Height of triangle through chromatographic peak
HK	Hexokinase
HPLC	High-performance liquid chromatography
I	Intensity of transmitted radiation; current (electrochemistry); nuclear spin quantum number (nuclear magnetic resonance spectroscopy)
I_0	Intensity of incident radiation
ΔI	McReynold's constant
i.d.	Internal diameter
IE	Ion-exchange
IEF	Isoelectric focusing
IR	Infrared
ISE	Ion-selective electrode
J	Coupling constant
k	Partition ratio (gas chromatography, high-performance liquid chromatography); Boltzmann constant
k'	Capacity factor
K	Partition coefficient
K_a	Acid dissociation constant
K_w	Dissociation constant for water
l	Pathlength
λ	Wavelength (lambda)
λ_b	Blaze wavelength
λ_{max}	Wavelength of maximum absorption
Λ	Molar conductivity (lambda)
LC	Liquid chromatography
m	Mass (mass spectrometry); magnetic quantum number (nuclear magnetic resonance spectrometry) order (ultraviolet)
M	Molar concentration
M_x	Magnetization along the x-axis
M^+	Molecular ion
M_D	Mass of daughter ion
M_F	Mass of charged fragment
M_S	Angular momentum quantum number
μ	Nuclear magnetic moment (mu)
MS	Mass spectrometry
n	Non-bonding (ultraviolet); theoretical plate number (gas chromatography, high-performance liquid chromatography)
N	Normality
n_{eff}	Effective number of theoretical plates
$NADP^+$	Nicotinamide adenine dinucleotide phosphate
NADPH	Nicotinamide adenine dinucleotide phosphate (reduced form)
NMR	Nuclear magnetic resonance spectroscopy
ν	Frequency (nu)
ν_R	Resonance frequency for reference
ν_S	Resonance frequency for sample
$\bar{\nu}$	Wave number
ODS	Octadecylsilyl
pH	$-\log_{10} a_{H^+}$
pI	Isoelectric point
π^*	Pi antibonding

PGB	Prostaglandin B
pK_a	$-\log_{10}K_a$
PMR	Proton magnetic resonance
ppm	Parts per million
p.s.i.	Pounds per square inch
Q	Molecular charge
r	Internal radius (gas chromatography, high-performance liquid chromatography); radius of arc of deflection (mass spectrometry)
R	Resistance (electrochemistry); gas constant
R_F	Retardation factor
rf	Radiofrequency
R_s	Resolution
RI	Refractive index
rmm	Relative molecular mass
RP	Reversed phase
S_l	Magnetization signal of liquid
S_s	Magnetization signal of solid
SCE	Saturated calomel electrode
SCOT	Surface-coated open tubular
SDS	Sodium dodecylsulphate
SEC	Size-exclusion chromatography
SHE	Standard hydrogen electrode
σ	Shielding constant (nuclear magnetic resonance spectroscopy) (sigma)
σ^*	Sigma antibonding
SIM	Selected ion monitoring
STP	Standard temperature and pressure
T	Absolute temperature
T_1	Spin–lattice relaxation time constant
T_2	Spin–spin relaxation time constant
t_p	Time of pulse
t_0	Retention time for non-retained component
t'	Adjusted retention time
t_r	Retention time
τ	Fluorescence lifetime (tau)
TD	Thermionic detector
TLC	Thin-layer chromatography
u	Velocity
$\bar{u}$	Mean linear velocity
UV	Ultraviolet
v	Velocity of ion
V	Voltage
V_0	Interstitial volume
V_p	Pore volume
V_r	Retention volume
w_b	Peak width at base
W_b	Width of triangle at base
$w_{1/2}$	Peak width at half-height
WCOT	Wall-coated open tubular
x', y', z'	Axes in rotating coordinate system
z	Charge

SI units

SI units have been used throughout this book, except for a few instances where other units are still in common use. The following tables are provided to assist the reader in the use of SI units.

Table 1 Basic SI units

Physical Quantity	*Unit*	*Symbol*
Length	metre	m
Mass	kilogram	kg
Time	second	s
Electric current	ampere	A
Temperature	kelvin	K
Amount of substance	mole	mol

Table 2 Derived SI units

Physical Quantity	*Unit*	*Symbol*	*Definition of unit*
Energy	joule	J	$kg\,m^2\,s^{-2}$
Force	newton	N	$kg\,m\,s^{-2} = J\,m^{-1}$
Power	watt	W	$kg\,m^2\,s^{-3} = J\,s^{-1}$
Electric charge	coulomb	C	$A\,s$
Potential difference	volt	V	$kg\,m^2\,s^{-3}\,A^{-1} = W\,A^{-1}$
Resistance	ohm	Ω	$kg\,m^2\,s^{-3}\,A^{-2} = V\,A^{-1}$
Magnetic flux	weber	Wb	$kg\,m^2\,s^{-2}\,A^{-1} = V\,s$
Magnetic flux density	tesla	T	$kg\,s^{-2}\,A^{-1} = Wb\,m^{-2}$
Frequency	hertz	Hz	s^{-1}
Pressure	pascal	Pa	$kg\,m^{-1}\,s^{-2} = N\,m^{-2}$
Conductance	siemens	S	$A^2\,s^3\,kg^{-1}\,m^{-2} = \Omega^{-1}$

Table 3 Non-SI units used in this book

Physical quantity	*Unit*	*Symbol*	*Conversion to SI unit*
Pressure	bar	bar	1×10^5 Pa
	pound per square inch	p.s.i.	$703.07\,kg\,m^{-2}$
	torr	Torr	101 325/760 Pa
Length	foot	ft	3.048×10^{-1} m
Magnetic induction	gauss	G	1×10^{-4} T
Electrical energy	electron volt	eV	$\sim 1.602 \times 10^{-19}$ J

1 Introduction to instrumental methods of analysis

There can be no doubt that the nature and range of chemical analyses that are feasible has changed dramatically over the last two decades. Increasing demand for more detailed analyses at lower levels has led to the development of a vast range of sensitive instrumental techniques. The analyst is now in a position to provide analytical data of a precision and accuracy, even on trace components, that would not have been contemplated 20 years ago. This armoury of techniques has been exploited in all areas of the biological sciences. The impetus for increased sensitivity and precision has come from a number of sources. For example, public concern over the quality of our environment and food has placed a considerable onus on regulatory bodies to monitor a vast range of materials, from river water to processed foods, for many hundreds of organic and inorganic compounds. The increasing sophistication of medical diagnosis has also made demands on the detection and quantification of many compounds in biological samples such as blood, plasma and urine. The analyst is therefore faced with demands for the detection and quantification of many analytes and the situation is further complicated by the wide range of matrices in which these compounds are to be found, from the relatively simple, such as drinking water, to the extremely complex, for example, faeces.

The usefulness of an analytical method, whether instrumental or not, must be assessed by the consideration of several criteria which must all be taken into account when decisions as to the preferred method are made. These criteria may also be used to illustrate the advantages of instrumental methods.

1.1 Precision and accuracy

In general terms, techniques which rely on objective measurements are more likely to provide precise data than those that are operator-dependent (subjective). This is one of the main advantages of instrumental techniques, but nonetheless an estimate of the precision of a method should always be made by replicate analyses, quoting the resulting variability as the coefficient of variation. The vast majority of instrumental techniques provide quantitative data by a comparison of standards and samples under identical conditions. The accuracy of the method then depends on the accuracy of the standard, the precision of the method and the validity of the assumption that the analyte in the standards and in the sample are being measured in exactly the same way. In other words, there must be no interference in the samples from other

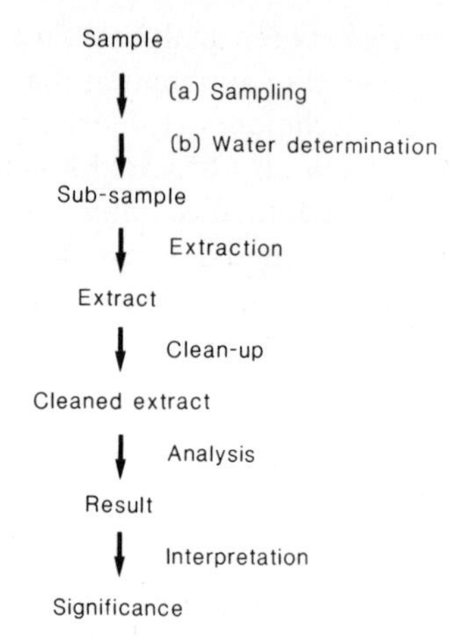

Figure 1.1 Typical analytical scheme. Water determination is often required so that data can be reported on a dry-weight basis.

compounds or matrix material. The accuracy of an instrumental method is therefore dependent on its specificity; the more specific a method, the less likely it is to be subject to interference. It is very easy with sophisticated instruments to produce precise but wholly inaccurate data due to undetected interferences. Extraction losses will also contribute to this inaccuracy (see Figure 1.1). Standards for precision and accuracy must be established before a method is accepted.

The validity of a method is often checked by comparison with existing accepted methodology. The fact that two independent methods produce the same results does not necessarily mean that these results are accurate; however, if they show different results, then clearly something is wrong with at least one of them. New methods may also be checked by arranging collaborative trials, where a number of laboratories undertake the same analyses on identical samples using the prescribed method and subsequently compare their results. Different laboratories will only obtain similar results, even when using the same methodology, if the instruments themselves are accurate. For example, a method based on the determination of the absorbance of a solution at a designated wavelength cannot possibly give consistent results if the wavelength or absorbance calibrations of the various instruments are not correct. Thus, the validation of instruments in terms of their specification becomes of paramount importance and instrumental calibration should be an important part of any setting-up routine.

1.2 Speed of analysis

The time taken for an analysis can be important if the results are required urgently, and time also contributes to the cost of the analysis. In the latter case, a distinction needs to be made between analytical time and operator time and this brings into question the level of automation that can be used. A further consideration is that some techniques, for example thin-layer chromatography (TLC), can analyse a number of samples simultaneously compared with techniques such as high-performance liquid chromatography (HPLC), and so the analysis time *per sample* should be evaluated. The instrumental part of the analysis may be the least time-consuming part and the time taken for extraction and clean-up must also be taken into consideration.

1.3 Cost

The cost of the analysis may be broken down into three parts:

(i) the cost of consumables (reagents, glassware, etc.)
(ii) the cost of the equipment
(iii) the staffing cost.

The cost of consumables per sample analysed is easily calculated and is often the smallest contribution to the overall cost. However, the cost of equipment for modern instrumental techniques can be extremely high; for example an automated HPLC system could cost £30 000 (at 1986 prices), while a mass spectrometer will be nearer £200 000 (1986 prices), depending on the sophistication of the data-handling required. These vast sums of money need to be considered in relation to the number of analyses that are likely to be completed by the instrument within its working life. The capital cost per analysis may then be seen as a more realistic sum, especially if the instrument is well utilized with a high degree of automation in sample preparation and analysis. The staffing cost will depend on the operator time required per analysis and the level of expertise required to run and maintain the instrument. Here again, the level of automation will have a direct influence over the cost of the analysis, as, once it is set up correctly, only relatively unskilled assistants may be required.

1.4 Safety

The safety of any laboratory procedure is of paramount importance. The use of instrumental methods of analysis, as opposed to traditional 'wet-chemical' methods, does not significantly reduce the hazards of toxic and caustic chemicals in the laboratory, as these chemicals will often still be required in the preliminary stages of sample preparation. However, care should be taken to ensure that the instrumental methods do not introduce further hazards and, where potentially dangerous situations are unavoidable, for example in high fluid pressures in HPLC or high electric potentials in mass spectrometry, it is essential that these are brought to the attention of the operators.

1.5 Automation

Some of the advantages of automation, such as reproducibility, increased system utilization and reduced operator time, have already been mentioned and it is certainly the suitability of many instruments for automation that has increased their widespread use. Thus, liquid or gas chromatographs can be fitted with autosamplers, as can spectrometers or even atomic absorption instruments. Furthermore, complete sets of experimental parameters can be stored in memory, or on magnetic disc, so that the instrument can be rapidly set up for any particular analysis. The actual control of the instrument is only one facet of the use of microprocessors in instrumentation; equally important is the ability to store and process the analytical data. The instrument can then be programmed to analyse a number of samples in sequence and to store the data. It is also possible, by designating certain samples as standards, to arrange for the microprocessor to calculate the final quantitative data. Several instruments may be linked to a central processor so that results from various analyses may be collated automatically.

The instrumental stage of the analysis is often only part of the overall method. A typical analytical scheme can be broken down into a number of stages as shown in Figure 1.1. Sampling is a crucial part of the analysis and extreme care must be taken to ensure that the sub-sample is representative of the original sample. The sampling scheme required will depend on the heterogeneity of the sample. In some cases, the next two stages of extraction and clean-up may not be necessary, but in general it is not possible to analyse the sample directly, either because the analyte is in an unsuitable form or there are too many interfering components present. In most analyses of compounds in biological samples it is these stages which require the most skill and contribute considerably to the overall error of the method as well as being time-consuming. Furthermore, it is these same stages that are extremely difficult to automate and this reduces the advantages of being able to automate the final instrumental stage of the analysis. At present, much development work is in progress in the area of sample preparation, including multidimensional chromatographic techniques and the use of robotics to automate simple manipulations. It must be remembered that the validity of the final data is dependent on the accuracy of each stage of the analysis, including efficient extraction and clean-up, and it is in these stages that most errors are to be found.

Instrumental techniques of analysis may be broadly classified into non-separative and separative, as shown in Table 1.1. In the former, the specificity of the determination relies on measuring some physical property which is specific to the component of interest in a sample or extract. Thus, if fluorescence is to be used to determine mycotoxins in an agricultural product, it will only produce accurate data if the mycotoxins are the only compounds in that sample that fluoresce at the analytical wavelength. Clearly in very

Table 1.1 Classification of instrumental techniques

Separative	*Non-separative*
Liquid chromatography	Refractometry
Gas chromatography	Hydrometry
Electrophoresis	Ultraviolet, infrared and visible spectrophotometry
Dialysis	Fluorometry
Reverse osmosis	Nuclear magnetic resonance spectrometry
Ultrafiltration	Mass spectrometry
	Atomic absorption spectrometry
	Polarimetry
	Optical rotatory dispersion
	Circular dichroism
	Raman spectroscopy
	Radiochemical techniques
	Sedimentation

complex mixtures non-separative techniques are liable to interference, that is to say there will often be other components present which respond in a similar manner to the compound of interest. In theory, these problems may be overcome by using separative techniques, where the compounds of interest are completely separated from the sample matrix, before identification and quantification. In practice, complete separation may not be possible or may be extremely difficult to achieve, and in these cases a partial separation may be carried out followed by a specific instrumental analytical technique.

The more important instrumental techniques, both non-separative and separative, as applied to the biological sciences, are discussed in this book. It has not been possible to cover all the relevant instrumental techniques within a book of this size and where omissions have been necessary, references to alternative texts are given. Throughout the chapters emphasis has been placed on the fundamental principles of the techniques involved, together with some examples of their more important areas of application.

Recommended general texts

Bassett, J., Denney, R.C., Jeffery, G.H. and Menham, J. (1983) *Vogel's Textbook of Quantitative Inorganic Analysis*, 4th edn, Longman, London.

Ramette, R.W. (1981) *Chemical Equilibrium and Analysis*, Addison-Wesley, Reading, MA.

Skoog, D.A. (1985) *Principles of Instrumental Analysis*, 3rd edn, Saunders College Publishing, Philadelphia.

Willard, H.H., Merritt, L.L., Dean, J.A. and Settle, F.A. (1981) *Instrumental Methods of Analysis*, 6th edn, Wadsworth, Belmont, CA.

2. Liquid chromatography

2.1 Introduction

Chromatography, in its simplest form, may be described as a process that allows resolution of a mixture of compounds as a consequence of the different rates at which they move through a stationary phase, under the influence of a mobile phase. This general definition does not restrict the nature of the phases involved and indeed it is the wide range of such phases that makes the techniques of chromatography so diverse. For example, the mobile phase could be a gas passing over a stationary liquid phase, leading to gas–liquid chromatography, or both the mobile and stationary phases could be liquids leading to liquid–liquid partition chromatography. This chapter is concerned only with those techniques in which liquid mobile phases are employed; those techniques employing gas as the mobile phase are discussed in Chapter 3.

The earliest demonstration of chromatography used a liquid mobile phase and is accredited to Day (1897–1903) and Tswett (1906) who separated plant pigments on columns of chalk. It was also this initial separation of coloured materials that led to the term chromatography (writing in colour). Relatively little progress was then made until the introduction of thin-layer techniques by Izmailov and Shraiber (1938), in which the stationary phase was held as a layer on a flat plate. This allowed simple development and compounds could be easily visualized, with spray reagents if necessary. The major development was, however, undoubtedly the work of Martin and Synge (1941) for which Martin received the Nobel prize in 1952. During the period 1940–1950 these workers laid down much of the fundamental theory of the chromatographic process and used it to predict the necessary developments to improve chromatographic resolution. It was this basic work which led to the introduction of gas chromatography with its inherent advantages of improved sample transfer between phases and hence resolution. Also, these early papers postulated the criteria necessary for high resolution liquid chromatography, that is to say, small particle sizes of large surface area, with associated high solvent pressures. The bases for gas chromatography and high-resolution liquid chromatography were thus laid some 30 to 40 years ago, yet it was gas chromatography that developed into a routine instrumental technique in the early 1950s and not liquid chromatography. The main reasons for this uneven development were essentially technological, in that the techniques for transferring liquids under high pressures at reliable flow rates were simply not available. However, it was realized subsequently that a liquid chromatographic technique, with similar resolving power, would be advantageous for many types of compounds, especially those that are non-volatile (such as ionic compounds, amino acids, dyes, macromolecules) or those that are thermally unstable (for example unsaturated lipids or compounds prone to decomposition, oxidation or polymerization). The last decade has therefore witnessed an enormous increase in interest in instrumental liquid chromatographic techniques, especially high-performance liquid chromatography (HPLC).

A very wide range of liquid chromatographic techniques exists, from simple column chromatography through thin-layer techniques to HPLC. The situation is further complicated by the large number of modes of chromatography available, that is to say the mechanisms by which the solute molecules (compound on interest) transfer between phases, for example in adsorption and ion-exchange. However, all the modes of chromatography may not be suitable for use with all the various techniques, for example size-exclusion on thin layers is not very common. Despite this complexity it is still possible to discuss a generalized theory of liquid chromatography which will apply to all the various techniques and modes, and this must be the starting point for a sound understanding of chromatography.

2.2 Theory of liquid chromatography

The theory of liquid chromatography may be most simply discussed for column chromatography, in which the stationary phase is held in a vertical column and the liquid mobile phase flows over it. However, the theory can be readily applied to all techniques.

The basic chromatographic process consists of partition of the sample

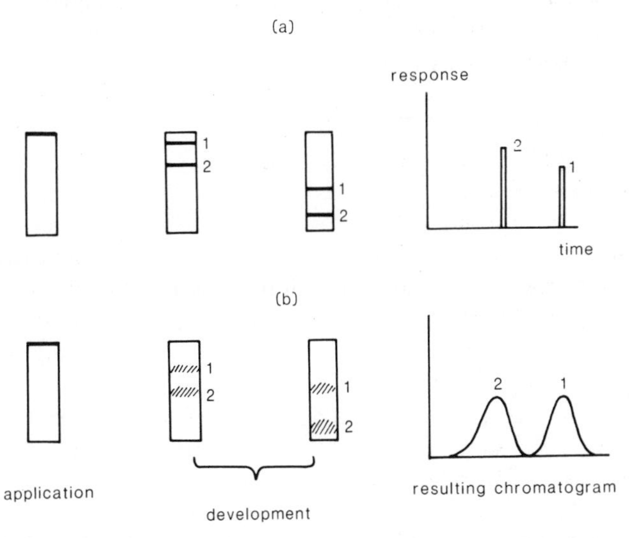

Figure 2.1 (*a*) Ideal chromatographic separation; (*b*) actual chromatographic separation with separation (spreading) within bands.

molecules between the two phases. While the sample molecules are in the mobile phase they will travel down the column and while they are associated with the stationary phase they will not. Thus, if the sample consists of two or more types of molecules, these will only be separated if they partition between the phases to different extents. An ideal separation between the components of a binary mixture is shown in Figure 2.1(*a*). In this case the components have separated but there is no separation within the components, in other words like molecules have stayed together and unlike molecules have separated. In practice, however, this situation is never achieved and the process depicted in Figure 2.1(*b*) will take place. Clearly if the separation within a band becomes similar to the separation between bands then resolution has been lost. This statement can be defined more precisely by the term resolution R_s, where

$$R_s = \frac{2(t_{r_1} - t_{r_2})}{W_{b_1} + W_{b_2}} \quad (2.1)$$

where t_{r_1}, t_{r_2} are the retention times and W_{b_1}, W_{b_2} are the mean peak widths of bands 1 and 2 respectively. With reference to Figure 2.2(*a*), the term $(t_{r_1} - t_{r_2})$ can be seen as a measure of the separation between bands and $(W_{b_1} + W_{b_2})/2$, the mean peak width, is a measure of the separation within the bands. Now the resolution may be improved by increasing the former or reducing the latter and it must be remembered that the entire value of chromatography as an

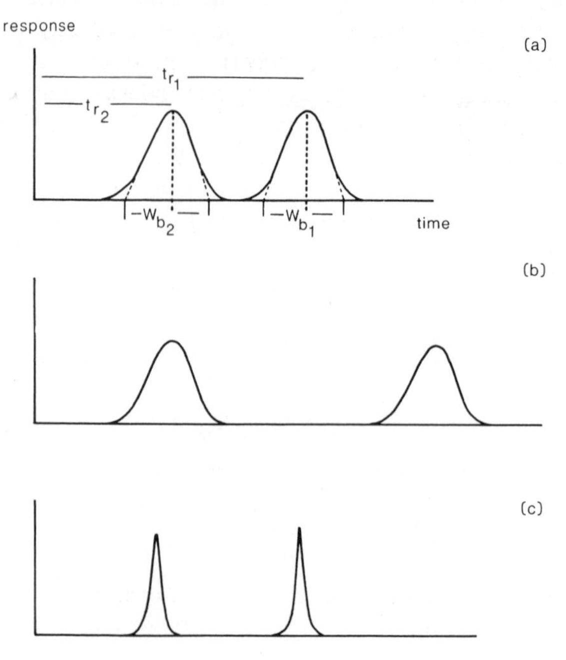

Figure 2.2 (*a*) Chromatographic resolution; increased (*b*) by increasing $(t_{r_1} - t_{r_2})$; or (*c*) by decreasing $\frac{1}{2}(W_{b_1} + W_{b_2})$.

analytical technique is based on its ability to resolve, and hence differentiate, closely related compounds.

The difference between retention times depends directly on the differential partition between the phases, which is thermodynamically controlled, whereas the band width depends on the efficiency of transfer of solute molecules within and between the phases and is therefore kinetically controlled. In general, it is advantageous to increase the resolution by minimizing band widths, as this does not increase the analysis time (Figure 2.2). These two aspects of the chromatographic process are essentially independent and may therefore be conveniently discussed separately.

2.2.1 *Chromatographic retention*

The rate of passage of a component down a chromatographic column can be related to the solvent velocity and the mean fraction of time spent by molecules of that component in the mobile phase R. Referring to Figure 2.1, if u is the velocity of the mobile phase and u_1 is the velocity of component 1, then these must be related by R, that is,

$$u_1 = uR \tag{2.2}$$

If the sample molecules partition completely into the mobile phase (that is, $R = 1$), then they will pass down the column at the same rate as the mobile phase itself. This corresponds to zero retention, often designated as t_0. Conversely, if the sample molecules partition completely into the stationary phase (that is $R = 0$), the component will be completely retained and will never be eluted from the column. Clearly, for successful chromatography, R must be between 0 and 1 and must be different for each component. If the partitioning between the two phases is considered as a dynamic equilibrium, the ratio of the amounts (moles) of solute in the phases must be constant. This ratio is called the capacity factor k' and

$$k' = \frac{N_s}{N_m} \tag{2.3}$$

or

$$k' + 1 = \frac{N_s}{N_m} + \frac{N_m}{N_m}$$

$$= \frac{N_s + N_m}{N_m}$$

Furthermore, the value of R can be expressed in terms of the amounts of the component in the two phases:

$$R = \frac{N_m}{N_s + N_m} \tag{2.4}$$

from which

$$R = \frac{1}{1 + k'}$$

Substituting for R from (2.2):

$$u_1 = \frac{u}{1 + k'} \tag{2.5}$$

The velocity of a component down a chromatographic column cannot be *directly* recorded and it is more usual to determine the time taken for it to be eluted, that is, the retention time, t_r. Clearly, u and t_r are simply related:

$$t_{r_1} = \frac{L}{u_1} \quad \text{and} \quad t_0 = \frac{L}{u}$$

where L is the column length and t_0 is the time taken for a component with zero retention to emerge from it. Equation (2.5) may then be re-formed by substituting for u_1 and u and eliminating L:

$$t_{r_1} = t_0(1 + k')$$

or

$$k' = \frac{t_{r_1} - t_0}{t_0} \tag{2.6}$$

The retention time of a component is therefore dependent on k', which is simply the ratio of the amounts of material in the two phases. Now these amounts (moles) can be expressed in terms of concentrations X in the respective phases and their volumes V, that is, $N_m = X_m V_m$ and $N_s = X_s V_s$. Substituting back into (2.3):

$$k' = \frac{K V_s}{V_m} \tag{2.7}$$

K is simply the distribution coefficient which governs the equilibrium model. It can also be seen that the capacity factor k' will be influenced by the ratio of the phase volumes (phase ratio β); the greater the proportion of stationary phase available for partition, the greater the retention. In addition to the phase ratio, the capacity factor k', and hence the retention, will be altered by any parameter which affects the distribution coefficient. The first of these, and probably the most important, is the nature of the phases themselves. In contrast to gas chromatography (Chapter 3), the mobile phase is an important parameter in the equilibrium and indeed to achieve changes in retention it is more usual to keep the same stationary phase and to alter the eluting power of the mobile phase. The exact nature of the required change will depend on the mode of chromatography involved (see Section 2.3). Only when such changes in the mobile phase have failed to achieve the required degree of retention would the stationary phase be altered.

The expected degree of retention from the equilibrium model will only be realized when the dynamic situation approximates closely to the theoretical ideal. In situations where this is not possible, deviations will occur. The simplest example is when the sample size is increased to such an extent that 'equilibrium' is no longer possible and the column is said to be overloaded. This results in a decrease in retention beyond a particular sample size (known as the linear capacity). Such an effect is also associated with a decrease in resolution as the bands broaden (see also Section 2.2.2).

The equilibrium process is thermodynamically controlled and thus will be sensitive to changes in temperature, provided there is a change in energy associated with the partition changes, which in this case corresponds to a change from solution in the mobile phase to absorption by the stationary phase. In liquid chromatography this energy change (ΔH) is much smaller than in gas–liquid chromatography (GLC) and therefore the associated temperature effects are much smaller. Nonetheless, even modest temperature changes, such as $\pm 10\,^{\circ}C$, can produce significant changes in retention times and lead to problems of both peak assignment and quantification when peak heights are used as these vary with retention. Thus, temperature control does become important, but is unlikely to be used as a means of altering retention in practice.

2.2.2 *Band broadening*

The second factor contributing to the observed resolution between components is their peak widths, strictly mean peak width, and this is the reason for the non-ideal separation shown in Figure 2.1(*b*). The final observed peak width is the result of contributions from several independent processes taking place within the column. These processes are kinetic in nature and may be broadly divided into those dependent on diffusion and those arising from mass transfer. In the former group, two mechanisms may be identified; simple diffusion and eddy diffusion. Any compound dissolved in a solvent, in this case the mobile phase, will tend to diffuse so as to eliminate differences in concentration. Any such diffusion will be both across the diameter of the column, which will not contribute to band broadening, and along the column (longitudinal diffusion), which will. The extent of longitudinal diffusion, and hence its effect on band broadening, depends on the diffusion coefficient of the compound in the mobile phase and also its velocity; the slower the mobile phase, the greater the extent of diffusion. Eddy diffusion is a direct consequence of the diverse pathways that sample molecules may take through the column. Where the solvent pathway is wide, the flow will be fast and hence the sample molecules will pass rapidly down the column. In narrower pathways, however, with slower flows, the molecules will move less rapidly. The extent of eddy diffusion depends primarily on the particle size of the stationary phase material and also on how well the particles are packed to form a homogeneous bed.

The mass-transfer terms are somewhat more complex and result from the transfer of solute molecules either within the mobile phase or within the chromatographically active layer on the stationary phase, the latter effect leading to delayed transfer between the phases and hence band broadening. In modern chromatographic materials, the stationary phase is synthesized to allow very efficient transfer between the phases and so this effect is quite small. Across any solvent pathway there will be a velocity gradient with the fastest stream in the centre and therefore, as sample molecules transfer across this pathway, they will be moved at varying rates down the column, leading to broadening. An extreme case would be when there is a pocket of stagnant mobile phase, as in a deep pore, and inefficient and variable transfer in and out of these sites will lead to considerable broadening. These contributions depend on the particle size, the solvent velocity and also on the diffusion coefficient in the solvent.

The various contributions can be combined, although not simply added, to produce an overall expression for the band broadening (Poole and Schuette, 1984). This is usually expressed in terms of the height equivalent to a theoretical plate, H, which is a measure of column efficiency:

$$H = L/n$$

where L is the column length and n is the number of theoretical plates for the column; n reflects the column's ability to produce narrow peaks, that is, its performance. The term originally arose from the number of plates required to achieve a given separation in a distillation column. Referring to Figure 2.2(a), the theoretical plate number if defined as

$$n = 16\left(\frac{t_{r_1}}{t_{W_{b_1}}}\right)^2$$

It should be noted that different solutes will provide different values for n on the same column. The relative importance of the contributions to band broadening, and hence H, vary with the chromatographic conditions and their effects may be summarized as follows:

(a) H is smaller (that is more efficient columns) in stationary phases with small particle size.
(b) H is smaller at low solvent velocities (but increases at very low values).
(c) H is smaller for solvents with high diffusion coefficients. The diffusion coefficient of a solvent depends on its viscosity, which in turn is temperature-dependent. Therefore more efficient chromatography is obtained at higher temperatures.
(d) H is smaller for small molecules, due to their more rapid diffusion rates.

These effects may be conveniently shown in diagrammatic form (Figure 2.3). Here it can be seen that small particle sizes lead to more efficient columns and furthermore that this may be achieved at higher solvent velocities, resulting in

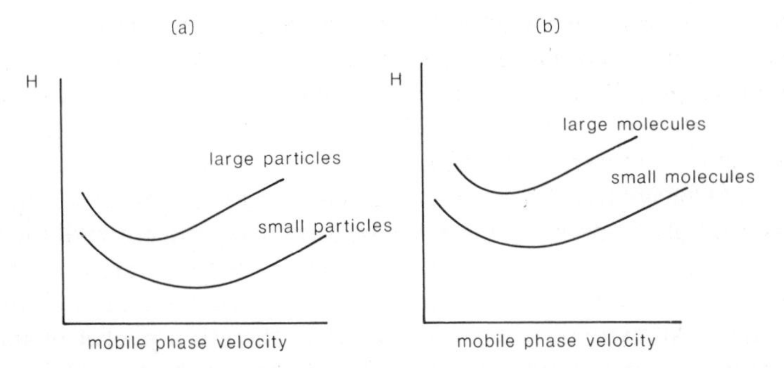

Figure 2.3 Effect of mobile phase velocity on column efficiency for (*a*) large- and small-particle stationary phases, and (*b*) large and small sample molecules. Note that small values of *H* represent efficient columns.

reduced analysis times. The introduction of small-particle stationary phases has been the single most important development in liquid chromatography both in terms of increased resolving power and on its effect on the associated instrumental development. Figure 2.3(*b*) simply shows the effect of reduced diffusion rates of large molecules in the solvent leading to less efficient chromatography and also that lower solvent velocities are required to counteract this effect.

The theory then allows us to predict those conditions most suitable for efficient chromatography and how changes in operating parameters will affect our results.

2.3 Modes of liquid chromatography

The mode of chromatography used for a particular separation refers to the mechanism by which the sample molecules interact with the stationary *and* mobile phases. It is important to note that the mobile phase plays an active role in the separation process. The theory outlined above can be applied to all the modes available, although the contributions of the various factors to band broadening may differ slightly. The five major modes are shown in Table 2.1, together with the basis for separation.

Table 2.1 Modes of chromatography

Mode	*Basis of separation*
Adsorption	Polarity
Partition	Solubility
Ion exchange	Charge
Size exclusion, Gel filtration, Gel permeation	Molecular size
Affinity	Biological activity

These will be discussed in turn as the factors affecting retention and resolution differ in each case.

2.3.1 *Adsorption chromatography*

Adsorption chromatography was the first mode of chromatography used and in simple terms involves the partition of solute molecules between solution in the mobile phase and adsorption on to the solid stationary phase. For this reason, the mode is also often known as liquid–solid chromatography (LSC).

The most commonly encountered stationary phase for adsorption chromatography is silica, although alumina has also been used, particularly in those cases where the slightly acidic nature of the surface silanol groups of silica may lead to decomposition of solute molecules. The surface of the silica gel is hydrated under normal conditions and the exact extent of this hydration has a significant effect on its chromatographic properties. The process of activation, commonly carried out in TLC by drying the prepared plates at 110–120 °C, removes only the outer layers of hydration, and more firmly bound water molecules are retained. However, it is these outer layers of water molecules that are important chromatographically and which modify the affinity between solute molecules and stationary phase.

The simple mechanism of adsorption outlined above is not completely satisfactory in explaining all the observed features of adsorption chromatography. A more reasonable interpretation of events can be obtained by also considering the interactions of solvent molecules with the stationary phase (Figure 2.4). The resulting solvated silica can then interact with solute molecules in one of two ways. The solute may become associated with the solvated surface when the mechanism involved is more similar to partition between two liquid phases. Alternatively, the solute molecules may displace the solvent molecules and become directly associated with the surface of the silica, namely true adsorption. Which of the two mechanisms actually operates in a given case will depend on the relative polarities (see below) of the solvent and solute molecules. For example, when a very polar solvent is used as a

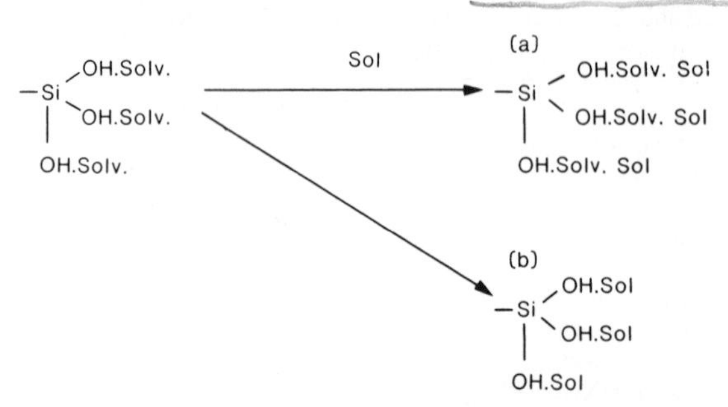

Figure 2.4 Interaction of solute molecules (Sol) with solvated surface of silica. (*a*) Molecules leading to association (partition); (*b*) more polar molecules leading to displacement (adsorption).

modifier in the mobile phase, the solute molecules will interact with the adsorbed monolayer of this solvent and the displacement mechanism will only occur with solutes of higher polarity.

The concept of a solvated surface for the silica is useful as it helps explain phenomena which the simple adsorption/desorption theory cannot.

In addition to the state of solvation of the stationary phase surface, the degree of retention of the solute molecules will depend on the strength or polarity of the mobile phase. Solvent strength is often defined in terms of ε° values, which are the heats of adsorption per unit area of the solvent; the higher the ε° value, the greater the eluting power of the solvent. The values are not identical for different stationary phases, but they follow a similar order (Table 2.2). In column chromatography, an increase in ε° value of 0·05 will result in a decrease of k' by a factor of 3–4. Mobile phases of any desired strength can be obtained from binary mixtures of suitable composition. Mobile-phase solvents may be characterized further by selectivity, which is a reflection of their specific interactions with solute molecules. Thus, solvent systems may be devised having the same strength but completely different affinities for different solute molecules. The practical significance of this is that complex mixtures of solute molecules of differing structures, but possibly with similar polarities, may still be separated by taking advantage of solvents of different selectivity. A detailed discussion of selectivity parameters is beyond the scope of this book and interested readers are referred to standard texts (see 'Further reading', p. 40).

The importance of the degree of hydration of the stationary-phase surface has already been stressed and it is therefore not surprising that traces of water in the mobile phase will similarly have a significant effect on retention. However, traces of water often also have a dramatic effect on peak shape, in particular, reducing tailing and hence improving resolution. The water deactivates the stronger adsorption sites leading to a more uniform and reproducible stationary phase. However, the problems of standardizing the degree of activation of stationary phases in adsorption chromatography have

Table 2.2 Elution strengths of solvents commonly used in adsorption chromatography

	Solvent strength ε°	
Solvent	*Silica*	*Alumina*
Hexane	0·01	0·01
Chloroform	0·26	0·40
Methylene chloride (Dichloromethane)	0·32	0·42
Ethyl acetate	0·38	0·58
Tetrahydrofuran	0·44	0·57
Acetonitrile	0·50	0·65
Methanol	0·7	0·95

not been totally overcome and for this reason bonded-phase chromatography (see Section 2.3.2) has become far more important.

2.3.2 *Liquid–liquid partition chromatography*

Partition chromatography, strictly liquid–liquid partition chromatography, may be considered as an extension of adsorption chromatography, the major difference being that instead of the silica-gel surface being solvated, it is now coated with a heavy loading of the required stationary liquid phase. The degree of retention of a solute molecule by such a phase depends directly on its relative solubility in the two phases and hence a good indication of a solute's chromatographic behaviour may be obtained from its partition coefficient between the two liquid phases in question. Clearly these phases must be immiscible in normal liquid–liquid extraction techniques and the same applies to partition chromatography. This then leads to the use of a non-polar phase with a relatively polar phase, and it is possible to use either of these as the stationary phase. In the normal mode, the silica gel would be coated with the polar phase, whereas the use of the non-polar phase as the stationary phase would lead to reversed-phase chromatography. In normal-phase elution, the solute molecules will be eluted from the column in order of increasing polarity. This is because the more polar molecules will have a greater affinity for the polar stationary phase, in exactly the same manner as in adsorption chromatography. However, with reversed-phase chromatography, where the stationary phase is now hydrophobic, the solute molecules will be eluted in *decreasing* order of polarity. Furthermore, the eluting power of the mobile phase will increase with *decreasing* polarity. Reversed-phase chromatography has now developed into the single most important form of liquid chromatography especially for HPLC, and finds wide application in the analysis of biological samples.

The stationary-phase support material is most commonly silica gel and the polar surface silanol groups provide excellent means of retaining polar liquid phases (that is, for normal elution). For hydrophobic liquid phases (reversed-phase), however, the surface of the silica must be modified to provide a means of bonding and this is most easily carried out by reacting the surface silanol groups with a modifying agent such as trimethylchlorosilane. The usual method of coating the support material, modified or not, is by solvent evaporation. The liquid stationary phase is dissolved in a volatile solvent and the dried silica gel added. The solvent is then removed by evaporation with constant agitation to ensure a uniform coating of the silica gel. This is most simply carried out in a rotary evaporator. The coated material is then packed in a column for use. Techniques are also available for *in-situ* coating of the silica gel after packing into a column, but these are less reliable.

The major advantage of liquid–liquid partition chromatography is that the choice of phases is infinite, and specialized phases can be produced for specific applications. However, the stationary phases are only weakly held to the support material and there is a tendency for them to be stripped off by the mobile phase, resulting in a rapid deterioration in performance. One solution to this problem is to chemically bond the stationary liquid phase to the support material so, provided that the bonding involved is compatible with the mobile phase, no loss of stationary phase will be experienced. The original bonded phases were prepared by esterification with alcohols to form silicate esters (≡Si—OR). However, these are not chemically inert and are readily hydrolysed by aqueous mobile phases, restricting their use to non-aqueous solvents. The majority of present-day bonded phases are, in fact, of the siloxane type (≡Si—O—Si—C≡) prepared either with chlorosilanes or alkoxysilanes. The characteristics of phases prepared by these reagents depend on the functionality of the chlorosilane; monochlorosilanes leading to monomeric coatings and di- and trichlorosilanes leading to polymeric phases containing much higher levels of the required stationary phase. However, if the degree of polymerization is excessive, the phase may have poor mass-transfer characteristics leading to poor peak shape. It is very difficult to obtain complete coverage of the surface silanol groups and these may lead to mixed mechanisms operating during elution. For this reason it is becoming usual to 'cap' any residual hydroxy groups with a reactive chlorosilane, such as trimethylchlorosilane, although this is not possible for phases with certain functional groups, such as amino-phases.

Bonded phases prepared in this manner may be polar or non-polar depending on the nature of the bonding species and a wide range of types exist in both categories. The most commonly encountered polar phase is probably the aminopropyl phase, prepared using triethoxyaminopropylsilane. This material is widely used for the analysis of polar water-soluble molecules, such as sugars, with aqueous alcohols (or more commonly acetonitrile) as the mobile phase. This is an example of normal-phase chromatography and so compounds are eluted in order of increasing polarity. An increase in eluting power of the mobile phase is also associated with an increase in polarity, in this example an increase in the proportion of water in the aqueous solvent. Thus, if gradient elution were to be used, the proportion of water in the mobile phase would also be increased during the chromatographic run.

Hydrophobic bonded phases in reversed-phase chromatography are rapidly becoming the most widely used modes of liquid chromatography. Initially, applications were confined to column techniques, but its influence is spreading to others such as TLC. The C_{18} bonded phase (also known as octadecylsilyl or ODS) is by far the most commonly used, often in conjunction with mobile phases of aqueous alcohols or aqueous acetonitrile. This combination alone can handle a very large number of compounds of differing polarity, simply by changing the mobile-phase composition; high concentrations of alcohols are used for the relatively non-polar (that is, hydrophobic) compounds and high concentrations of water for the more polar solutes which are *less* well retained.

In a mixture of components of widely differing polarity a solvent gradient could be advantageously used, as discussed in Section 2.4.4.

When very hydrophobic compounds are studied, it may be necessary to use a more powerful (that is, more hydrophobic) solvent for elution, such as acetone, chloroform or even hexane. This type of chromatography is known as non-aqueous reversed-phase chromatography and is extensively used in lipid analyses, for example, in the separation of triacylglycerols. Alternatively, the degree of retention can be altered by changing the hydrophobicity of the stationary phase, for example instead of C_{18} a C_8 or even a C_2 bonded phase could be used, though clearly this is not as simple as changing the mobile phase, and the latter would be attempted in the first instance.

Liquid–liquid partition chromatography has been traditionally associated with the separation of small molecules (relative molecular masses < 1500), while larger molecules would be handled with size-exclusion techniques (Section 2.3.4). However, in recent years, partition phases, in particular reversed-phase systems, have been developed for use with macromolecules such as proteins. This has been possible for two main reasons, firstly the development of phases with superior mass-transfer characteristics, even for large molecules, and secondly the use of larger-pore silica as the base support material, which allows better access to the chromatographic sites.

The versatility of reversed-phase chromatography may be further extended to charged molecules, which are highly polar and are therefore poorly retained on a hydrophobic stationary phase. However, by removing their net charge, they form more hydrophobic species. This may be accomplished by ion-suppression, which simply involves shifting the pH of the mobile phase such that the ionization is reduced. For an acid, this means that the pH should be significantly (1–2 pH units) below its pK_a. An alternative approach is the formation of ion pairs of the charged solute molecules with a hydrophobic counter-ion. The resulting 'ion pair' will be more hydrophobic and will therefore be more strongly retained on the column. The degree of retention can then be altered to suit the analysis in question by altering the counter-ion used.

$$RCO_2^- + (CH_3)_3N^{+}\sim\sim \rightleftharpoons RCO_2^- N^{+}(CH_3)_3\sim\sim\sim$$

Polar solute ion	Counter-ion	Relatively hydrophobic ion pair

In reality, the situation on the column is more complex, but the above mechanism does allow a simple explanation of observed chromatographic effects. For more theoretical texts, see 'Further reading'.

2.3.3 *Ion-exchange chromatography*

The interaction between solute molecules and the stationary phase in ion-exchange chromatography is based on charge difference, so that the molecules and exchanger must carry complementary charges. The ion-exchange process may be described in two stages: adsorption and desorption. In the first stage, the charged solute ion P^- is attracted to a site with opposite charge on the exchanger R^+ displacing a counter-ion X^-:

$$(R^+X^-) + P^- \rightleftharpoons (R^+P^-) + X^-$$

In the second stage, the solute ion is displaced by a competing salt ion S^-:

$$(R^+P^-) + S^- \rightleftharpoons (R^+S^-) + P^-$$

This second stage represents elution of the solute from the chromatographic column.

There are two types of ion exchanger based on the sign of the charge they carry and hence the type of ion that can be retained. An anion exchanger will have a net positive charge and a cation exchanger, a negative charge. These materials may be further classified in terms of how their charges vary with changes in pH, in an analogous manner to strong and weak acids. Indeed, these same expressions are used. Examples of the various types of resin are shown in Table 2.3. The net charge on an ion exchanger, which is a measure of its ability to retain oppositely charged ions, will clearly depend on the pH of the mobile phase with which it is in contact. Furthermore, the net charge will depend on the pK_a of the ion-exchange group involved. The relative capacity, that is, the charge, of a weak anion and cation exchanger are shown in Figure 2.5. Clearly, carboxymethyl (CM) exchangers would not function as cation exchangers below pH 3 or 4 and diethylaminoethyl (DEAE) exchangers would not function adequately as anion exchangers above pH 11 or 12. The situation with strong exchangers is less complex, as extremes of pH would have

Table 2.3 Types of ion exchanger

Strength	*Anion exchanger*	*Cation exchanger*
Strong	Quaternary ammonium salts $—^+NR_3$	Sulphonic acids $—SO_3^-$
Weak	Tertiary amines $—C_2H_4—\overset{+}{N}H(C_2H_5)—C_2H_5$	Carboxylic acids $—CH_2CO_2^-$

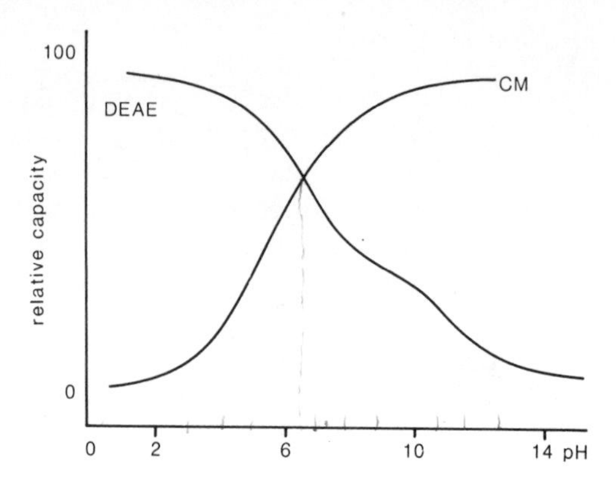

Figure 2.5 The effect of pH on the charge of weak cation exchanger (CM) and a weak anion exchanger (DEAE).

to be employed to suppress their ionization, conditions which are unlikely to be encountered in practice. For example, the pH of the mobile phase would have to be below 1 to suppress effectively the ionization of a strong cation exchanger such as a sulphonic-acid based resin.

The influence of pH on the charge of the stationary phase is only one side of the equation, because for the ion-exchange process to take place, the solute molecules must carry a complementary charge. Here again, the pK_a's of the ionizable groups will dictate the effect of pH. For some compounds, for example, salts of strong acids, pH will be relatively unimportant, whereas for molecules containing both potential anionic and cationic groups, such as amino acids or proteins, changes in pH can alter the net charge significantly, even to the extent of changes in sign. This is important as it allows proteins and similar compounds to be separated by either anion or cation exchange, simply by altering their net charge by pH shifts. If the iso-electric point of the protein is known, that is, the pH at which it has zero net charge, this may be used as a starting point in choosing the pH conditions for adsorption.

Once the required charge conditions have been obtained for both exchanger and solute molecules (ions) and adsorption has been achieved, the second stage (desorption) can be studied. This is usually carried out by the addition of competing ions to the mobile phase to displace the charged solute molecules from the exchanger. In certain cases, ion exchange may be used as a technique solely to isolate charged molecules from solution, and therefore the desorption process need not be selective and high concentrations of salt ions may be used. However, it is more usual to use ion exchange to separate several charged compounds, as for example in amino-acid analysis. In this case, selective desorption is required so that different ions are desorbed at different rates and are therefore separated. This effect is shown in Figure 2.6 for three compounds adsorbed onto an ion exchanger. The three compounds have different affinities

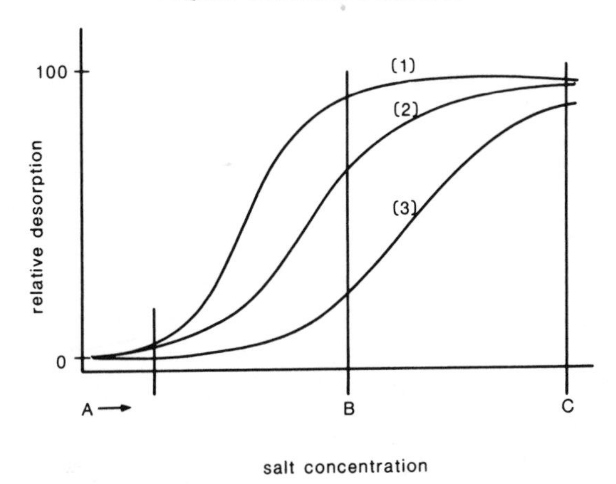

Figure 2.6 The effect of salt concentration on the desorption of three components (1), (2), (3) on an ion exchanger. *A*, complete retention; *B*, selective desorption, separation; *C*, complete desorption, elution but *no* separation.

for the exchanger and can therefore be separated by correct selection of the salt concentration. At a salt concentration *A*, none of the components is desorbed to any extent and so they remain on the column, whereas at the salt concentration *C* they are all virtually desorbed and elute rapidly, but with no separation. At some intermediate concentration *B*, the rates of desorption are sufficiently different for separation to be effected.

Charged molecules may also be eluted from exchangers by altering the pH of the mobile phase so that the proportion of solute molecules, or ion-exchange sites on the exchanger, that carry the correct charge is reduced, as shown in Figure 2.5. This in turn will reduce the degree of retention, ultimately allowing elution. In a similar manner to salt desorption, care must be taken to select pH changes that will allow selective desorption and hence separation. It is also possible to combine the effects of salt desorption with shifts in pH of the mobile phase to permit the separation of complex mixtures. This procedure is often encountered in amino-acid analysis.

In many experimental situations there may be no indication as to the necessary mobile-phase composition to allow selective desorption, either in terms of salt concentration or pH. One method of establishing suitable conditions is to form a gradient in the mobile phase, of either parameter (salt or pH) and observe the approximate conditions under which the components of interest elute. A suitable mobile-phase composition may then be estimated. Another situation where gradient elution may be helpful is where the sample to be analysed contains a wide range of components in terms of their affinities for the exchanger. Under isocratic conditions of elution this would result in very long analysis times with very broad peaks for the more highly retained components. A pH or salt gradient would allow an increase in the eluting power of the mobile phase during the chromatographic run, resulting in narrow peaks even for those compounds with very high adsorption affinities (see also Section 2.4.4).

A wide range of materials have been used to form ion exchangers, including synthetic resins of cross-linked polystyrene, cross-linked polydextrans, cellulose and silica. In each case, the ion-exchange functionality is chemically bonded to the surface. In order for the materials to be used as small particles under high pressure (to improve resolution) they must be rigid. This involves the use of highly cross-linked polymers, which may adversely affect access to the ion-exchange sites, or the use of naturally rigid materials such as silica. Synthetic resins, such as cross-linked polystyrene, may also exhibit secondary retention effects due to hydrophobic bonding. Thus, two proteins with identical charge characteristics but with different hydrophobicities may still separate due to the increased retention of the more hydrophobic component. This also means that the addition of an organic solvent to the eluting buffer may influence retention and/or resolution.

Silica-gel based ion exchangers are not widely used, mainly because of their poor stability at extremes of pH, especially in alkaline media.

A branch of ion-exchange chromatography, often discussed as a separate topic, is that involving the separation of simple inorganic cations and anions, known as ion chromatography. The principles involved are identical to those outlined above and so should be considered as simply another application area of ion exchange. The major difference is in the instrumentation used, particularly the techniques available for detecting solute ions of interest in the column eluate, a problem which is compounded by the presence of buffer ions. Thus, if conductivity measurements are to be used to detect eluting ions, either the buffer ions must be removed by an additional ion-exchange column or the background conductivity must be removed electronically. In certain cases, the ions of interest may have other properties, for example, ultraviolet chromophores, and this will facilitate their detection.

2.3.4 *Size-exclusion chromatography*

The terms size exclusion, gel filtration and gel permeation now appear to be used interchangeably to describe those techniques capable of separating molecules on the basis of differences in molecular size as shown in their ability to penetrate pores within the stationary-phase material to differing extents. The principle of the separation mechanism is illustrated in Figure 2.7. Large molecules will not be able to enter the pores and so will be eluted in the interstitial volume V_0 (that is, the volume of mobile phase between particles), whereas smaller molecules will be able to enter the pores to differing extents depending on their size. In the extreme case, small molecules will be able to enter all of the pore volume V_p and so will be eluted with a volume of $(V_0 + V_p)$.

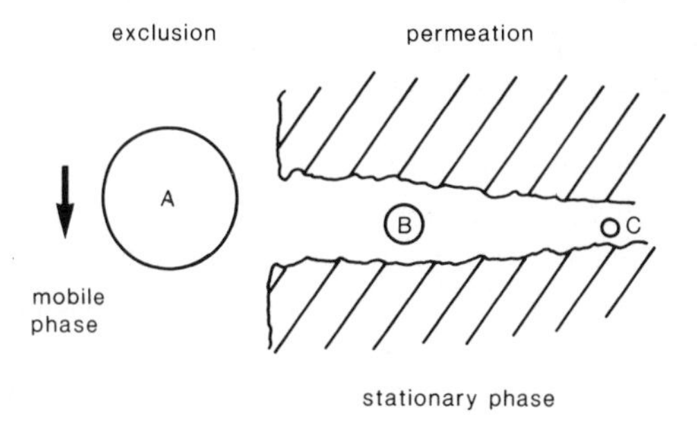

Figure 2.7 Mechanism of size exclusion. *A*, Complete exclusion; *B*, partial penetration; *C*, complete penetration.

In general, the retention volume may be expressed by

$$V_r = V_0 + K_D V_p$$

where K_D is the distribution coefficient, which represents the degree of penetration into the pores. If the process is not complicated by secondary effects, then separation will be achieved solely on the basis of molecular size. However, it should be remembered that this cannot be directly related to relative molecular mass as molecules of similar mass but different shape will be able to penetrate the pores to different extents and so will be separated. The molecular-mass fractionation range, that is to say the range of molecular masses between complete exclusion and complete penetration, will therefore often be different for dissimilar types of molecules (for example, proteins or polysaccharides) on the same stationary phase, on account of their different shapes. A wide range of pore sizes is available in gel materials to provide columns with different fractionation ranges. In order to obtain the maximum resolution possible, a stationary-phase material with the lowest possible exclusion limit should be used. If an attempt is made to separate a group of small molecules (for example, of relative molecular masses 500–1000) on a phase with a high exclusion limit (for example, of relative molecular masses 10^6), they would all coelute as a single peak. There is a simple logarithmic relationship between relative molecular mass and elution volume, and calibration curves such as that shown in Figure 2.8 may be used to ascertain the molecular mass of unknown compounds after chromatography under identical conditions, provided they are of similar type.

The resolution afforded by size exclusion is generally poor compared with other chromatographic modes, especially for macromolecules with poor diffusion characteristics. However, it is a 'mild technique' with, at least in theory, no direct interaction between the solute molecules and the stationary phase. This leads to retention of biological activity of sensitive macromolecules, such as enzymes, as no denaturation takes place. Thus, any loss of

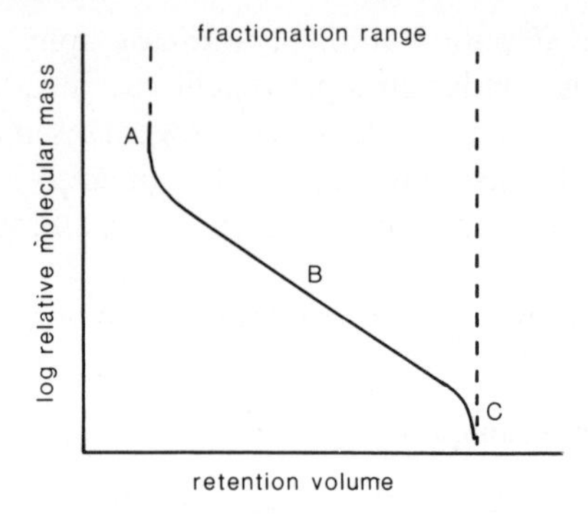

Figure 2.8 Calibration curve for size exclusion. *A*, Complete exclusion; *B*, partial penetration; *C*, complete penetration.

activity will be restricted to degradation in solution, which, as water is often used as the mobile phase, is small.

2.3.5 *Affinity chromatography*

The modes of chromatography outlined above are all based on general characteristics of the solute molecules, for example size or charge. However, it is possible to make use of the more specific interactions encountered in biological systems and this is known as affinity chromatography. The technique can be applied to almost any *reversibly* interacting pair of molecules with the most common applications coming from the following groups:

- antigen–antibody
- enzyme–substrate or inhibitor
- hormone–binding protein
- nucleic acid–complementary sequence

The 'partner' of the compound to be determined, or isolated, is immobilized, often by covalent bonding, to a support material and becomes the stationary phase. When the sample is applied to this material, any component that has suitable activity (binding capacity) will be retained and other compounds will be washed through. The mobile phase is then changed to reduce this interaction and elute the component, for example by shifts in pH or salt concentration. In extreme cases, it may be necessary to destroy the interaction by degradation of the solute with reagents such as urea or guanidine. In many cases the high degree of specificity of the interactions means that simple separations between retained and non-retained compounds give adequate purity for the latter. In other instances, where more than one type of molecule binds to the stationary phase, separation is still possible, provided selective desorption can be achieved by suitable choice of mobile phase, in an analogous manner to that described for ion exchange (Section 2.3.3). In some instances, the exact nature of the biospecific interaction may not be fully understood and this is exemplified in reactive dye affinity chromatography. A reactive dye such as Cibacron F3GA, is bonded to a gel which may then be used to purify a number of enzymes, especially kinases, dehydrogenases and other nucleotide-dependent species. It has been suggested that a 'dinucleotide fold' is the active site in the binding enzymes (Thompson *et al.* 1975), although other mechanisms may be involved. Further details on the chemistry of binding ligands to support materials may be found in a recent review by Calton (1984).

2.4 Chromatographic techniques

The various modes of chromatography described in Section 2.3 can be used in a number of different chromatographic techniques to effect separation of mixtures of components and their identification or isolation. For some of these techniques all the modes may be used, whereas for others, such as paper chromatography, the very nature of the technique dictates the retention mechanisms involved.

2.4.1 *Paper chromatography*

Paper chromatography (PC) is one of the simplest forms of chromatography and also one of the oldest. The stationary phase is simply a piece of filter paper with the mobile phase passing over it by capillary action (ascending chromatography) or a combination of capillary action and gravity (descending chromatography). Ordinary filter paper may be used, but chromatographic paper prepared from specially purified cellulose (98–99% α-cellulose) provides improved resolution and reproducibility. PC is most commonly used for the separation of hydrophilic compounds, such as amino acids, peptides and sugars, where the mechanism is primarily one of partition between the liquid mobile phase and the 'water–cellulose complex'. The separation process is, however, further complicated by secondary effects, such as adsorption to the cellulose and even ion exchange from the small number of carboxylic acid groups formed during the manufacture of the paper.

Reproducible results and good resolution will only be obtained if:

(a) the samples are applied as small discrete spots
(b) the sample solvent is evaporated prior to development
(c) the paper is equilibrated in the solvent vapour again prior to development, a process which may take in excess of one hour.

The spots may be detected by the same methods as used for TLC and similarly quantified. It should be noted that any spray reagent used must be compatible with the paper and this clearly excludes all those derivatizing reagents in strong acids or alkalis, for example, phenol/sulphuric acid for sugars.

PC has now very largely been replaced by TLC, where very similar separations can be carried out on plates coated with microcrystalline cellulose, but is still used in various biochemical fields, such as in 'protein mapping' where enzymic hydrolysates of proteins (peptides) can be separated, often on a preparative scale for further analysis, such as amino-acid composition. The versatility of PC may be increased by using impregnated papers such as silicone-oil treated materials for reversed-phase chromatography, but here again this may be more precisely carried out by TLC.

2.4.2 *Thin-layer chromatography*

In thin-layer chromatography, the stationary phase is a particulate material (particle size 5–30 μm, mean 20 μm) bound as a layer (0·1–0·5 mm thick) to an inert support, which was initially most commonly glass but is now more often plastic or aluminium. The plates are prepared, either commercially or in the laboratory, by coating the support with the stationary phase and a binder (gypsum, starch or polyesters) as a slurry in a suitable solvent and allowing the plates to dry, either under ambient conditions or in ovens. As shown in Section 2.2.2, the resolving power of a chromatographic system is very dependent on the particle size of the stationary phase, and in TLC a significant improvement in performance can be achieved by reducing the mean particle size to 5 μm with a narrow particle-size distribution. This development has led to the technique of high-performance thin-layer chromatography (HPTLC) by analogy with HPLC.

TLC is not restricted to particular modes of chromatography, provided a stable layer of the material of interest can be found. The most commonly encountered stationary phase is silica gel, which can be used to separate a very wide range of compounds in terms of differing polarity, simply by changing the solvent strength (Table 2.2). At one extreme, triacylglycerols can be separated with solvents such as hexane and diethyl ether and, at the other, polar species such as sulphonic-acid dyes with polar solvents such as aqueous alcohols. TLC plates are used only once so that any slight solubilization of the silica gel caused by using aqueous solvents at high pH (such as aqueous ammonia) may not matter, although a similar effect would not be acceptable with HPLC columns. A further material used for adsorption chromatography with TLC is alumina, which has many of the properties of silica gel, but unlike the latter tends to be basic rather than acidic. This difference can be important for the separation of compounds susceptible to acid- or base-catalysed degradation.

Microcrystalline cellulose may be used to provide the basis for a partition system which will perform in a similar manner to PC. The more uniform nature of the matrix in TLC results in improved resolution and also a reduction in the time required to achieve a given separation. Other modes of chromatography have been employed with TLC techniques, but only to a very limited extent. Thus, gel-filtration TLC has been used in the biochemical field for the separation of proteins and also for approximate molecular-mass determinations by comparison with standards of known molecular mass

(Miller 1978). Similarly, modified layers have been produced with ion-exchange functionality or with hydrophobic surfaces for reversed-phase applications.

Irrespective of the mode of separation, the technique is carried out in the same manner. The sample solution and relevant standards are spotted in a line about 1 cm from the base of the plate. The application solvent is then removed from the spot by evaporation prior to placing the plate in the solvent vapour *above* the mobile phase in a sealed tank which has previously been allowed to equilibrate. After equilibration of the plate this is lowered into the solvent for development by capillary action (only ascending chromatography is commonly used in TLC). In many instances, the equilibration stage is omitted, with the plate being placed immediately into the solvent after spotting and drying. The R_F (retardation factor) values (that is, the ratio of the distance the component moves to the distance the solvent front moves) are more likely to be variable under these conditions as the activity of the plate will not be constant but will depend on the drying conditions previously used. This applies in particular to plates coated with silica gel or alumina. Ideally, the separated components should form small symmetrical spots and, if poor spot shapes result, for example streaking behind the majority of the component, mixed mechanisms may be operating. An example of this kind of problem could arise from residual carboxy groups on cellulose. A similar effect is also observed when the plate is overloaded, and for preparative applications a narrow band is usually applied along the starting line, so that the sample is not concentrated on one small area of stationary phase.

2.4.2.1 *Visualization and quantification.* The majority of the compounds separated by TLC techniques are not coloured and therefore cannot be detected directly (obvious exceptions include synthetic dyes and natural pigments). If the compounds are fluorescent, then they may be detected simply by irradiation with ultraviolet (UV) light of suitable wavelength. Many compounds which contain a UV chromophore are able to quench fluorescence and so if a suitable fluorescence indicator is incorporated into the stationary phase (for example, manganese-activated zinc silicate), these compounds may be observed as dark spots on a fluorescent background. Any reagent that will react with the compounds of interest to form coloured derivatives can in theory be used as the basis of a spray reagent, with the restriction that the reaction conditions required must be compatible with the stationary phase (such as temperature, and acidity). The spray reagents may be very general, for example acidified dichromate will show up most organic compounds, or they may be much more specific, for example ninhydrin for amino acids. The use of highly specific reagents increases the 'identification power' of TLC in that if components have the same R_F values *and* produce the same reaction with a spray reagent, they are more likely to be the same compound.

TLC is usually associated with qualitative analysis, but it can be made into a quantitative technique of reasonable precision by a number of means. One simple procedure involves visual comparison between the separated component of interest and a range of quantitative standards of the same component. An *estimate* is then made as to the nearest standard in terms of spot intensity and size after visualization. This is really only a semi-quantitative method as the presence of other components in the sample may well affect the spot shape. Errors of $\pm$ 50% are not unreasonable and they may be even larger if errors in spotting are also significant.

A more precise method involves removal of the component from the TLC plate after elution. This is achieved by scraping off the relevant portion of the stationary phase and extracting the component with a suitable solvent, either in a batch-wise manner or in a Soxhlet extractor. The component may then be quantified in solution by UV or visible absorption, fluorescence or any other suitable technique. The preferred method of quantification, however, is based on *in-situ* reflectance measurements, as this avoids extraction losses and is far more rapid. In this technique, radiation of a defined wavelength (not necessarily monochromatic) is shone at the TLC plate, which may be moved across the beam (Figure 2.9). A photomultiplier is positioned so as to detect the reflected radiation, which will be reduced when a component absorbing at the wavelength(s) used passes across the beam. Standards and the sample are scanned under identical instrumental settings and quantification is achieved by comparison. It is also possible, on certain densitometers, to record absorption spectra of the components *in situ*, which assists identification.

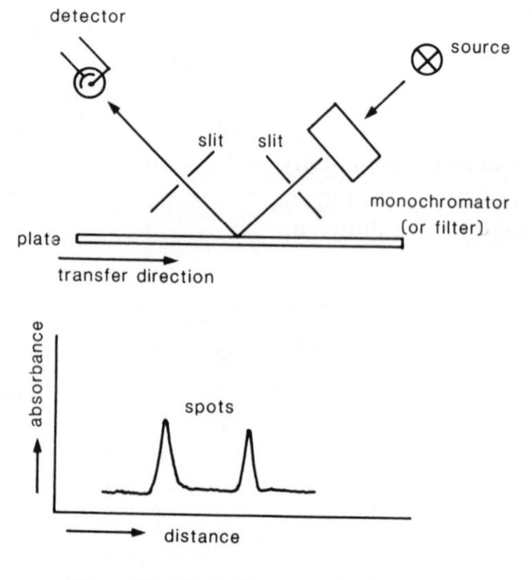

Figure 2.9 TLC reflectance densitometry.

2.4.3 *Column chromatography*

Open-column chromatography is traditionally carried out in glass columns with stationary phases of 50–80 μm diameter particles. The columns may be dry packed and then equilibrated with mobile phase or packed as a slurry. The sample is applied to the top of the column bed and the components are eluted with mobile phase which percolates through the column under the influence of gravity. The eluate is collected in fractions for subsequent analysis. There is no restriction on the modes of chromatography that can be used, and gradient elution can be carried out with a simple two-chamber mixing device. The basic technique may be improved by a number of modifications, for example, a uniform solvent flow may be achieved by placing a metering pump in the solvent line, either before or after the column. A continuous detector, similar to an HPLC detector (see Table 2.5) may be used to monitor the column eluate. The resolution of the technique may be improved by reducing the particle size of the stationary phase to 20–40 μm. This also reduces the solvent flow rate which may then be increased by the application of a modest (1–4 bar) pressure, often achieved by an inert gas over the solvent in the reservoir. This type of chromatography is often known as 'flash chromatography'.

In general, the resolution obtained by this type of column chromatography is relatively poor and applications are found mainly in sample preparation prior to other forms of analysis, as for example in sample clean-up prior to HPLC.

2.4.4 *High-performance liquid chromatography*

The chromatographic processes involved in HPLC are no different from those described in Section 2.2 and already encountered in the techniques described in sections 2.4.1–2.4.3. The complications arise from the more sophisticated instrumentation which is required to pump solvents under high pressures in a controlled manner and also from the more complex detectors used, although these are directly derived from many of the techniques described in this book. The major components of a simple isocratic (constant solvent composition) HPLC chromatograph are shown in Figure 2.10 and will be described in turn.

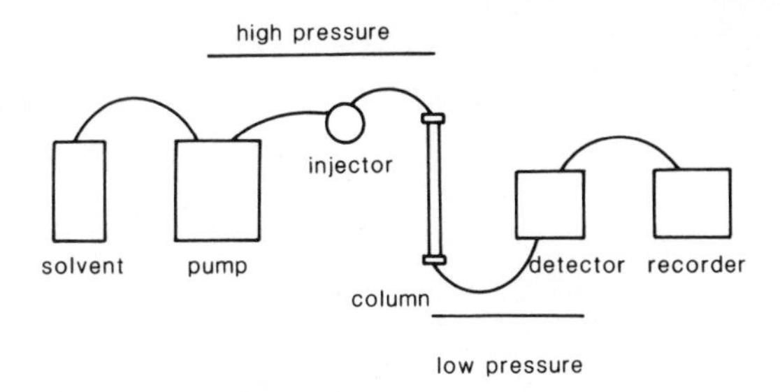

Figure 2.10 Basic high performance liquid chromatograph.

The solvent reservoir can be any convenient container, but the solvent itself must be degassed, by heating, applying vacuum or ultrasound, or by purging with helium. The solvent must also be free of particulate material and this is usually achieved with an in-line filter. The nature of the solvent used is clearly dictated by the mode of chromatography to be used and the particular compounds in the sample.

The pump is one of the most expensive components in the system (£1500–£3000 at 1986 prices) and its requirements depend on the dimensions of the columns to be used (Table 2.4). The most common analytical applications employ 4·6 mm internal diameter (i.d.) columns for which a flow rate of 1–2 ml min^{-1} is required. Most analytical pumps are constant-flow pumps based on a twin reciprocating-piston design, so that a relatively pulse-free flow is achieved. In single-piston pumps, the pressure, and hence flow, drops on the refill stroke and a pulse dampener is required to obtain a uniform output. Analytical pumps can often be modified for microbore work (low flow rates) or preparative work (high flow rates), but specialized pumps are also available for those applications. The reproducibility of the chromatograms produced is directly related to the reproducibility of the solvent flow and reliable pumping systems are an essential requirement of HPLC. Unfortunately, it is often the pumping systems which prove unreliable. Most pumps have a maximum working pressure of about 400 bar (6000 p.s.i.).

The sample, dissolved in a suitable solvent, is introduced onto the top of the column *via* an injection valve. The sample is initially injected into a holding loop which is then connected to the mobile phase stream by rotation of the valve. In this manner, the sample is introduced without disruption of the mobile phase; some dilution of the sample is inevitable but the convenience of these valves outweighs this slight disadvantage. Alternative sample introduction techniques such as 'stop-flow' and septum injection are now no longer widely used.

The chromatographic column is the 'heart' of the system where separation takes place. The most commonly used columns are 4·6 mm i.d. and 15–25 cm in length, packed with stationary phases of 5–10 μm particle size. However, as

Table 2.4 Pump flow rates

Type of chromatography	*Column dimensions i.d. × length* (mm) × (mm)	*Pump flow rate Typical values* (ml min^{-1})
Microbore	1 × 250	0·05
Microbore	2 × 250	0·20
Conventional	4·6 × 250	1·5
Fast	4·6 × 50 (3 μm phase)	4
Semi-preparative	8 × 250	10
Preparative	25 × 250	100

noted in Table 2.4, other configurations are possible and the advantages of microbore HPLC are now being realized, as efficient columns and reliable low-flow-rate pumps become available. Good resolution (that is, columns with high plate numbers) will only be realized when the columns are efficiently packed, and this involves slurry packing under high pressures. It is not a difficult procedure, but does demand suitable high-pressure vessels and strict safety control. Many chromatographers rely on commercially prepared columns. The nature of the stationary phase employed will depend on the nature of the compounds to be separated. All of the modes of chromatography described in Section 2.2 can be used in HPLC, even gel filtration since the advent of rigid porous materials. However, in the biological sciences, most of the HPLC separations routinely used rely on reversed-phase columns. This is due, in part, to the versatility of these columns and, perhaps more importantly, to their robustness and the ease with which they can be cleaned from contaminating components often found in 'dirty' biological samples.

The column eluate, with the components now separated, passes through the detector, where a response is generated in proportion to the amount of material passing through. The number of different detection systems is legion, based on both physical and chemical properties of the solutes; in addition, some techniques are based on the change of physical properties of the mobile phase, for example, refractive index (RI).

The UV detector, either as a fixed- or variable-wavelength instrument, is undoubtedly the most widely used detector in HPLC. In its simplest form, radiation of well-defined wavelengths is produced by interference filters (for example, 254 nm, 280 nm, etc.) and the absorbance of the eluate is monitored at these wavelengths.

The arrangement for a simple single-beam detector is shown in Figure 2.11. The design of the flow cell is critical so that it does not introduce dispersion (band broadening) into the system. The volume of the cell depends on the column dimensions but will be 5–10 μl for a 4·6 mm i.d. column and down to 1 μl or less for a 1 mm microbore column. Double-beam instruments are also available in which fluctuations in source output are automatically compensated for.

Clearly, the applicability of this detector is limited to those compounds which absorb appreciably at these wavelengths and the range of applications

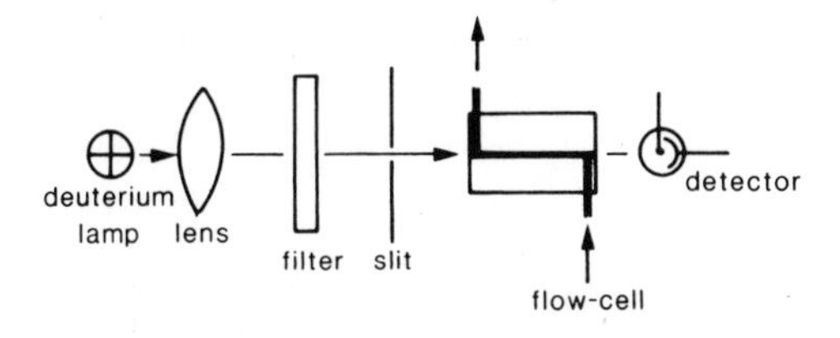

Figure 2.11 Single-beam filter ultraviolet detector.

can be extended by using a monochromator-based instrument which allows any wavelength to be selected, including visible wavelengths if a suitable source is available. It is not always the λ_{max} (the wavelength of maximum absorption) value of the compound of interest that should be used, as in certain cases a shift away from this wavelength may result in reduced interference from other compounds in the sample. In other words, a wavelength should be chosen which is a compromise between good sensitivity for the compounds of interest and good selectivity over the interfering compounds. Modern UV detectors have very high sensitivities, down to 0·001 absorbance units full scale (AUFS), which means that amounts of compounds of less than 1 ng can often be detected. Some detectors also have the ability to record spectra of the compounds, either by stopping the flow, so trapping the compound in the cell and then scanning, or by use of diode-array techniques in which the absorbance at many wavelengths is recorded simultaneously, from which the spectrum can be derived.

The fluorescence detector finds less application as an HPLC detector, simply because many fewer compounds are fluorescent than absorb in the UV or visible regions. This is both an advantage and a disadvantage, in the sense that the detector is now more specific and therefore less likely to be subject to interference by other compounds, yet on the other hand it can only be applied to more limited classes of compounds. Fluorescence detectors are extremely sensitive and therefore, in cases where they can be used, provide an excellent means of detection, with improved sensitivity and selectivity over UV absorption. The components of a simple filter-based fluorescence detector are shown in Figure 2.12; the versatility of the instrument can be improved by replacing one or both of the filters by monochromators. In this case, both the excitation and emission wavelengths can be adjusted to obtain the optimum compromise between high sensitivity and reduced interference. It should be noted that even if a *co-eluted* interferent is not fluorescent, that is, will not be detected, it may still affect the fluorescence of the analyte by quenching. The technique of standard additions (see Section 11.3.2) is very useful in situations such as this, as any added standard will be subject to the same quenching effect as the analyte in the sample.

The RI of the eluate from the column will change as components are eluted

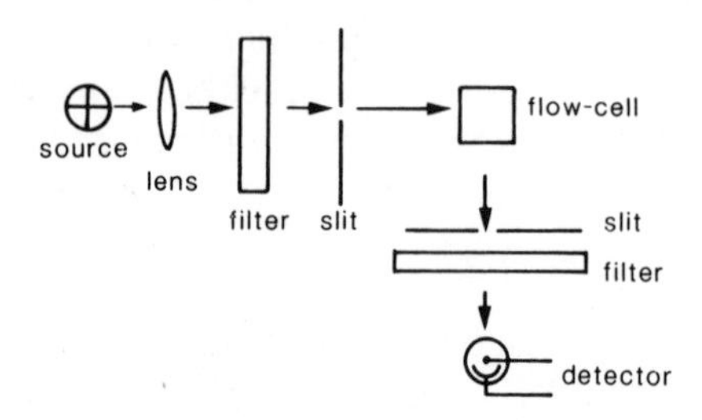

Figure 2.12 Filter fluorescence detector.

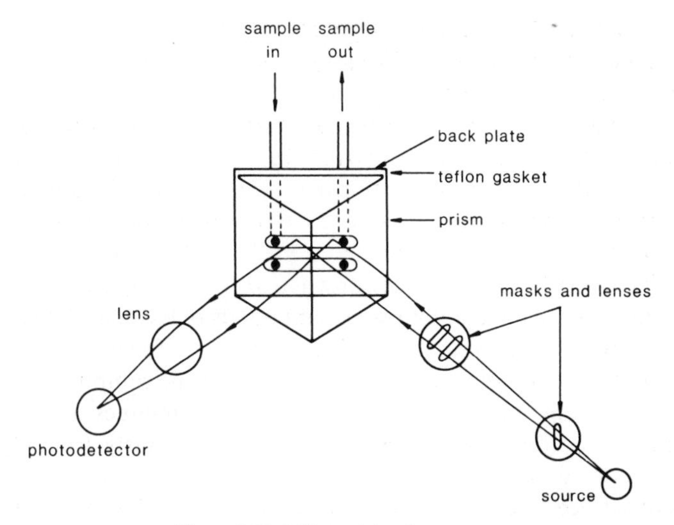

Figure 2.13 Differential refractometer.

and this provides the simple basis for a non-specific detector. Many optical designs have been employed but that based on the Fresnel prism seems to be the most widely used. As shown in Figure 2.13, it is a differential refractometer recording differences between the pure mobile phase, held in the reference cell, and the column eluate. Unfortunately, most RI detectors are not very sensitive to the presence of solutes, but they are very sensitive to changes in ambient conditions, such as pressure and temperature in the cell, and also to even very small changes in mobile-phase composition. This last factor rules out the use of RI detection with gradient elution. Certain instruments based on interference effects are considerably more sensitive, but still suffer from the other disadvantages mentioned above. It is not surprising to find that RI detection is only used in those cases where other detectors such as fluorescence or UV are not applicable, for example, with sugars and many lipid components.

Compounds that can be oxidized or reduced at a polarized electrode surface can be detected using the electrochemical detector (ECD). This is an extremely sensitive instrument for those compounds that undergo redox reactions at modest potentials and whose redox products do not poison the electrode surface. The ECD detector is rather similar to the RI detector in the sense that it is sensitive to changes in ambient conditions, particularly pressure, so that the solvent flow must be pulse-free. ECD has been widely used in the determination of very low levels of biologically active material, such as biogenic amines.

The more important characteristics of these detectors are summarized in Table 2.5, together with some of their areas of application. In particular, the excellent sensitivity and selectivity of fluorescence should be noted, and this is

Table 2.5 Comparison of HPLC detectors

Detector	*Sensitivity*	*Selectivity*	*Stability*	*Applications*
Ultraviolet/visible	Good	Good	Good	Most-used detector, e.g. drugs, vitamins, pigments
Fluorescence	Excellent	Excellent	Good	Vitamins, mycotoxins, many derivatized components, e.g. dansyl amino acids
Refractive index	Poor	None	Poor	Lipids, carbohydrates
Electrochemical	Excellent	Good	Poor	Catecholamines, ascorbic acid, sugars
Mass spectrometer	Good	Excellent	Interface problems	Any trace analysis, drugs, pesticide residues, etc.

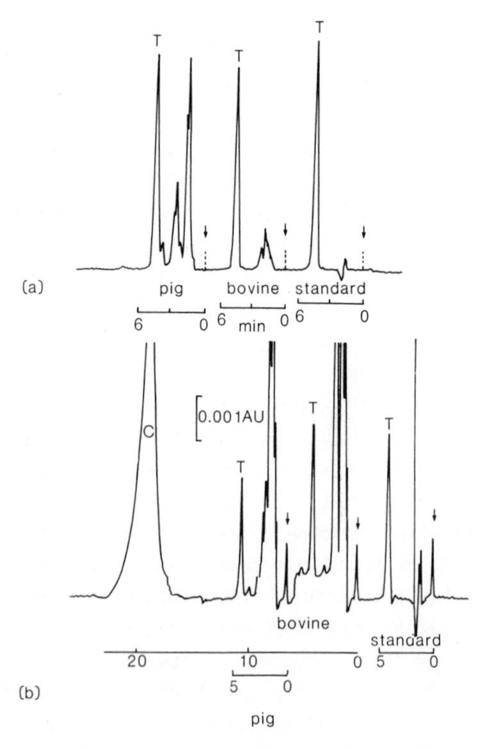

Figure 2.14 Comparison of (*a*) fluorescence and (*b*) ultraviolet detection of α-tocopherol in extracts of pig and bovine plasma. Chromatographic conditions: 30 × 0.39 cm stainless steel column packed with 10 μm μBondapak C18; eluent, methanol–water(97 : 3, v/v); flow rate, 3 ml min.$^{-1}$ Fluorescence detection λ_{exc} 296 nm; λ_{em} 330 nm; UV detection at 280 nm. Peak identification: T, α-tocopherol; C, β-carotene. Redrawn with permission from McMurray and Blanchflower (1979).

illustrated in Figure 2.14. The far simpler trace resulting from fluorescence compared with ultraviolet detection is a direct result of the greater selectivity of the former.

In addition to these four most commonly encountered detection systems, a wide range of more specialized instruments are available, ranging from infrared and conductivity detectors to mass spectrometers coupled directly to the chromatographic column. The latter is probably the nearest to a universal, yet specific and sensitive HPLC detector that has been achieved, but cost will preclude its widespread application.

In its simplest form, the output from the detector is fed to a chart recorder where the chromatogram is produced, that is to say, a trace of detector response as a function of time. The identity of compounds in the sample may be then tentatively assigned on the basis of retention times by comparison with standards. If the reproducibility of retention times is poor between chromatographic runs, the chromatographic similarity between the standard and a component may be confirmed by 'spiking' the sample with a portion of the standard. Further evidence of the structure of the compounds in the sample can be obtained by altering the chromatographic and/or detection conditions and observing if the standard and the sample component behave in the same way.

Once the identity of the compound of interest has been satisfactorily established, quantification can be carried out. This is simply achieved (external calibration) by comparison of peak heights (or preferably areas) of the quantitative standards and the corresponding peak in the sample. It is essential to demonstrate linearity by construction of a calibration plot, covering the expected range of concentrations of the component in the samples. The alternative techniques of standard additions (useful where interference is suspected) and internal standard (useful to compensate for errors in injection, although these are usually small with loop injectors) may also be used (see Section 11.2.2). Finding a suitable internal standard is not always easy, as it must be similar in terms of chemical properties and detectability to the compound of interest, but must elute in an area of the chromatogram free from interference. The basic liquid chromatograph shown in Figure 2.10 can be extended in a number of ways to increase the system's versatility and level of automation. The addition of a second pump with its controller allows simple binary-solvent compositions to be produced by setting the relative flow rates of the two pumps (Figure 2.15). Thus, if pump A is pumping solvent A at 0·5 ml min^{-1} and pump B is pumping solvent B at 1·5 ml min^{-1}, the total flow is 2·0 ml min^{-1} of composition 25% A and 75% B (V/V). A much more important function of the second pump and controller is that the solvent composition can be changed throughout a chromatographic run, that is, gradient elution can be carried out. If B is the more powerful solvent (the more polar solvent in adsorption chromatography or the more hydrophobic solvent in reversed-phase chromatography) then increasing the

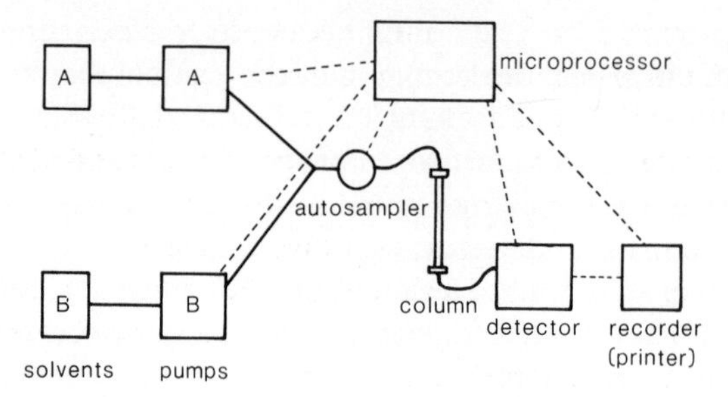

Figure 2.15 Microprocessor-controlled gradient chromatograph. ——— Solvent connections; ---- electrical connections.

percentage of B will increase the rate of elution of the components. This effect is illustrated in Figure 2.16 for reversed-phase chromatography of a complex protein mixture. With isocratic elution, the early eluting peak would not be resolved from the injection solvent and/or the later eluting peaks would be very broad and difficult to quantify. Both of these problems are solved by using a suitable gradient, in this case with increasing acetonitrile content. Gradient elution is invaluable for the separation of complex mixtures with widely varying polarities, as often encountered in biological systems. However, the technique should only be used where an isocratic separation is not possible, as

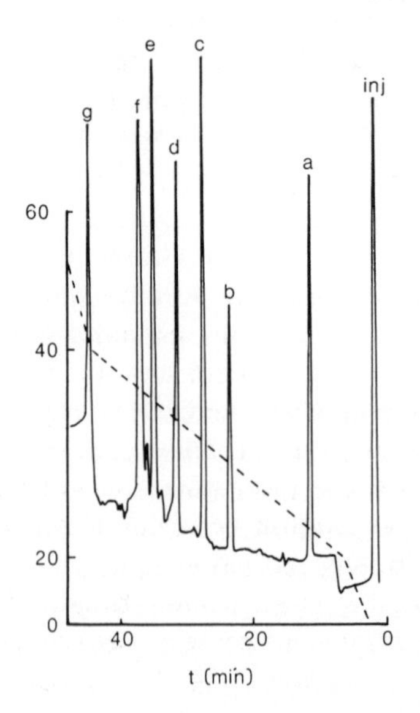

Figure 2.16 HPLC chromatogram of 6 proteins. Column: Hypersil-ODS (5 μm, 100 × 5 mm); mobile phase: primary solvent: 0.1 mol L^{-1} sodium dihydrogenphosphate/phosphoric acid (pH 2.1, total phosphate concentration 0.2 mol L^{-1}), secondary solvent: acetonitrile (the gradient profile is given by the dotted line); flow-rate: 1.5 ml min^{-1}; detection UV at 225 nm; *a*, tryptophan; *b*, cobra neurotoxin 3; *c*, ribonuclease A; *d*, insulin; *e*, cytochrome c; *f*, lysozyme; *g*, myoglobin. Redrawn with permission from O'Hare and Nice (1979).

quantitative precision is very often reduced. Furthermore, it must be remembered that time will also be required to re-equilibrate the column to the initial chromatographic conditions before subsequent analyses can take place. Solvent gradients can also be formed by proportional mixing before introduction to the pump. Here only one pump is required and the controller now governs proportionating valves on the solvent inlet lines.

The inclusion of an autosampler is essential for automatic operation of the chromatograph. The sampler can be set to inject a sample, taken in sequence from a rack, at a predetermined time or it may be coupled to a central processor so that the sample injection can be co-ordinated with the start of the gradient. The central microprocessor may also be used to control the detector, perhaps to change the wavelength of detection to suit a particular sample. In a completely automated system, it is unlikely that the detector response would simply be displayed on a chart recorder, but rather the data would be manipulated (peaks integrated, comparisons with standards made) so that final concentrations are reported.

2.5 Sample preparation

The nature of the samples encountered in many areas of the biological sciences means that extensive sample preparation is the norm rather than the exception. It is very rare that a sample can be taken and analysed directly, although there are a few exceptions, for example, synthetic dyes in soft drinks or trace metals in river water. Once an adequate sampling regime has been established, the sample preparation can be considered in two parts: extraction and clean-up.

2.5.1 *Extraction*

The aim of the extraction stage is to obtain the component of interest (analyte) in a suitable medium, often a solvent, and to leave the other undesired components behind. The extraction therefore should ideally be 100% efficient for the component of interest and have zero extraction efficiency for everything else. In practice, this ideal situation is very rarely achieved and so extraction becomes a compromise between efficiency and selectivity. This point can be illustrated by the extraction of sugars from dried foods or even dried blood. If water is used as an extractant, this would be very efficient for the sugars but would also extract some proteins, salts and pigments. Now if a less polar solvent is used, for example 80% (V/V) aqueous ethanol, the extraction of the sugars will still be adequate, but many of the interfering components will remain behind. The degree of selectivity required of the extraction, and indeed of the subsequent clean-up, depends on the specificity of the final analytical technique.

The choice of a suitable extracting solvent will depend on the solubility properties of the analyte and also on those of potentially interfering compounds. In all cases, once the choice has been made, recovery studies *must* be carried out to determine the percentage efficiency of the extraction. This is usually carried out by adding a known amount of the analyte to the sample and determining the percentage recovered after extraction. This value is compared with an unspiked sample. The recovery so determined will only be a measure of the amount of 'free analyte' in the sample. If any of the analyte is bound, for example to cell-wall material, it may still be 'missed' even if recovery studies are carried out. In such cases, attempts must be made to free the bound analyte before extraction, for example by acid or enzymic hydrolysis.

Rather than relying on differences in solubility in the extracting solvent to produce a clean extract, it is possible to use a more specific interaction to effect the isolation, for example ion exchange. This technique is widely used in the isolation of charged molecules, such as amino acids, vitamins and metabolites from physiological fluids, for example, urine. It may be necessary to adjust the pH of the sample so that the analyte has the desired charge characteristics for ion exchange to take place. Alternatively, the pH may be adjusted to suppress ionization to allow the analyte to be extracted into an organic solvent. This approach has been extensively used in the isolation of organic acids from plant materials.

An alternative approach to obtaining the analyte from the sample matrix is to destroy the latter, rather than to extract from it. In its simplest form, this involves combustion/pyrolysis of the sample leaving the analyte behind. Clearly this is only applicable to components that are stable at the elevated temperatures used (500–700 °C), which means metals and their salts, and even then the nature of the associated anion may well change, for example carbonates will degrade. A milder destruction of the organic matrix may be achieved by digestion, as opposed to ashing above, in which the sample is digested in boiling concentrated acids (sulphuric, nitric and even perchloric acids). The organic material is oxidized leaving metallic elements in solution ready for analysis, by atomic absorption for example. A modified form of digestion (also known as wet ashing) is used in the Kjeldahl procedure for determining nitrogen in food and other samples, where sulphuric acid is used with catalysts (Cu, Hg or Se) and the organic nitrogen is trapped as ammonium sulphate.

2.5.2 *Sample clean-up*

Judicious choice of extraction conditions may mean that the extract is sufficiently clean for direct analysis. However, in most cases, particularly when the analyte is present in only very small amounts, further sample preparation (clean-up) will be required. The range of such clean-up procedures is endless with every conceivable separative technique having found application in some area. In general, the procedures may be split into two groups, firstly those in which complete groups of unrelated compounds are removed and secondly those where compounds similar in nature to the analyte are removed.

2.5.2.1 *Removal of unrelated compounds.* Here advantage is taken of a difference in characteristics between the analyte and interfering compounds. Some of the possible group separations are shown in Table 2.6. In each of these examples some 'gross difference' between the analyte and interfering compounds is being exploited to yield a cleaner extract. In many ways this type of clean-up is very straightforward, but care must still be exercised to ensure that none of the analyte is lost, in other words the recovery of the clean-up step must also be evaluated.

2.5.2.2 *Removal of closely related compounds.* Many analyses may demand a more thorough clean-up, possibly in addition to the crude separations already described. This is often necessary where the analytical technique is relatively non-specific, or where the analyte is present in a matrix of similar compounds at much higher levels, which will tend to overload certain analytical techniques, for example HPLC or GLC.

Table 2.6 Examples of clean-up procedures

Basis	*Technique*	*Example*
Molecular size	Dialysis/reverse osmosis	Separation of lactose and salt from whey proteins
	Size exclusion	Separation of amino acids from proteins
Molecular charge	Ion exchange	Isolation of vitamins, e.g. thiamin
	Ion-pair formation	Isolation of organic acids, also sulphonic acid dyes
Solubility	Solvent extraction	Soxhlet extraction of lipids
		Solvent extraction of organic residues in potable/waste water
	Crystallization	Fractionation of sterols from other lipid components by crystallization from methanol
Polarity	Thin-layer chromatography	Isolation of tocopherols from non-saponifiable fraction of lipids on silica gel
		Isolation of drugs from urine
	Column chromatography	Isolation of vitamin D_2/D_3 prior to HPLC analysis
Volatility	Distillation	Collection of aroma or taint components prior to identification by other chromatographic techniques
Chemical stability	Saponification	Removal of triacylglycerols, by conversion to sodium salts of fatty acids, to allow examination of non-saponifiable fraction of oils and fats
	Oxidation	Destruction of sorbic acid in mixtures with benzoic acid (preservatives) using potassium permanganate prior to quantification of latter by UV absorption

The techniques employed in this type of clean-up are similar to those already discussed in Table 2.6, with more emphasis on those with greater separative power, such as the chromatographic methods. A relevant example would be the determination of carotene (α and β) in foods or plasma, where the analyte is separated from other lipid components by column chromatography on alumina, prior to quantification by UV absorption.

In most instances in the analysis of minor components in biological materials, using chromatographic techniques, it is in the sample-preparation stages that the problems arise and where the major sources of error are to be found. Recent developments, such as the use of disposable chromatographic cartridges, have greatly improved the clean-up stages in many analyses, but care must still be taken to ensure that losses of analyte do not take place and that potential interferents are removed.

Numerous texts have recently been published on the application of chromatographic techniques in the biological sciences, many of which contain details of sample preparation techniques (see Macrae 1982, Lawrence 1984, De Leenheer *et al.* 1985, Henschen *et al.* 1985).

References

Calton G.J. (1984) *Methods in Enzymology* **104**, 381.

Day, D.T. (1897) *Am. Phil. Soc.* **36**, 112.

Day, D.T. (1903) *Science* **17**, 1007.

De Leenheer, A.P., Lambert, W.E. and de Ruyter, M.G.M. (1985) *Modern Chromatographic Analysis of Vitamins*, Marcel Dekker, New York.

Henschen, A., Hupe, K-P., Lottspeich, F. and Voelter, W. (1985) *High Performance Liquid Chromatography in Biochemistry*, VCH Verlagsgesellschaft, Weinheim.

Izmailov, N.A. and Shraiber, M.S. (1938) *Farmatsiya*, **3**, 1.

Lawrence, J.F. (1984) *Food Constituents and Food Residues*, Marcel Dekker, New York.

Macrae, R. (1982) *HPLC in Food Analysis*, Academic Press, London.

McMurray, C.H. and Blanchflower, W.J. (1979) *J. Chromatogr.* **178**, 525.

Martin, A.J.P. and Synge, R.L.M. (1941) *Biochem. J.* **35**, 1358.

Miller, J.N. (1978) In *Chromatography of Synthetic Biological Polymers*, R. Epton (ed.), Ellis Horwood, Chichester, 181.

O'Hare, M.J. and Nice, E.C. (1979) *J. Chromatogr.* **171**, 209.

Poole, C.F. and Schuette, S.A. (1984) *Contemporary Practice of Chromatography*, Elsevier, Amsterdam, ch. 1.

Thompson, S.T., Cass, K.H. and Steilwagen, E. (1975) *Proc. Nat. Acad. Sci. USA* **72**, 669.

Tswett, M. (1906) *Ber. Dsch. Bot. Ges.* **24**, 384.

Further reading

Grob, R.L. (1983) *Chromatographic Analysis of the Environment*, 2nd edn, Marcel Dekker, New York.

Lurie, I.S. and Wittwer, J.D. (1983) *High Performance Liquid Chromatography in Forensic Chemistry*, Marcel Dekker, New York.

Munson, J.W. (1984) *Pharmaceutical Analysis*, Marcel Dekker, New York.

Provder, T. (1980) *Size Exclusion Chromatography*, American Chemical Society, Washington, DC.

Snyder, L.R. and Kirkland, J.J. (1979) *Introduction to Modern Liquid Chromatography*, 2nd edn. Wiley, New York.

3 Gas chromatography

3.1 Introduction

Gas chromatography (GC) achieves separation of mixtures by partition of components between a mobile gas phase and a stationary phase. The most common form of gas chromatography is gas–liquid chromatography (GLC) in which the stationary phase is an involatile liquid coated onto an inert solid support. The second type of gas chromatography is gas–solid chromatography (GSC) in which the stationary phase comprises particles of a solid adsorbent.

GLC was introduced by James and Martin (1952). In the early years, the vast majority of analyses were performed using packed columns in which the stationary phase is coated onto small particles of an inert solid before being packed into a coiled glass or metal column. In recent years, capillary columns have become widely used. The most common form of capillary column is the wall-coated open tubular (WCOT) column in which the stationary phase is coated onto the inside wall of a long glass or fused silica capillary.

GLC has become a major analytical technique which gives excellent separation of components from many complex mixtures. Over 200 stationary phases are available for use in GLC. These phases differ widely in chemical structure and consequently a suitable phase can often be found for the separation of quite complex mixtures. A wide range of detectors is also available, allowing the analyst to select a detector which yields high sensitivity or selectivity of detection appropriate to a particular analysis. The sample size required is extremely small (micrograms or less) and the analytical procedure can be automated allowing the routine analysis of large numbers of samples.

The major limitation of GLC is that the components must be stable and have significant volatility at the analytical temperature. However, the use of high temperatures allows the analysis of many compounds which would normally be considered relatively involatile. Mixtures of triacylglycerols, which contain components such as tristearoylglycerol having a vapour pressure of $< 0{\cdot}05$ mm at 300 °C, are commonly analysed by GLC. Simple derivatization reactions can often be used to convert involatile or reactive molecules, such as sugars, to stable compounds of sufficient volatility for analysis.

3.2 Principles

The time taken for a molecule to pass through a GLC column is known as the retention time t_r and is dependent on the partition coefficient K, where:

$$K = \frac{\text{mass of vapour dissolved in unit column length of stationary phase}}{\text{mass of vapour dissolved in unit column length of mobile phase}}$$

If $K = 0$ for a given molecule, the molecule is not retarded by the stationary phase, and the retention time t_0 represents the time taken to travel through the gas volume of the column. A molecule which is retarded by the stationary phase spends a time Kt_0 dissolved in the stationary phase, using the definition of the partition coefficient. Hence

$$t_r = t_0(1 + K)$$

The parameter t_r increases with an increase in column variables such as length, internal diameter, percentage of stationary phase coating the support and density of packing. It is reduced by an increase in the analytical temperature and is affected by the nature of the carrier gas. K is related to the partial vapour pressure of the component dissolved in the stationary phase. It is often useful to report the adjusted retention time t', where $t' = t_r - t_0$. The ratio of the adjusted retention times of two substances measured on a given column is independent of all variables except the temperature and the stationary phase, and therefore the ratio of the t' values for a known and unknown component can be useful for compound identification. An accurate value for the parameter t_0 can be determined by extrapolating the retention times of a homologous series of normal alkanes back to zero carbon atoms, although it is commonly estimated as the retention time of methane.

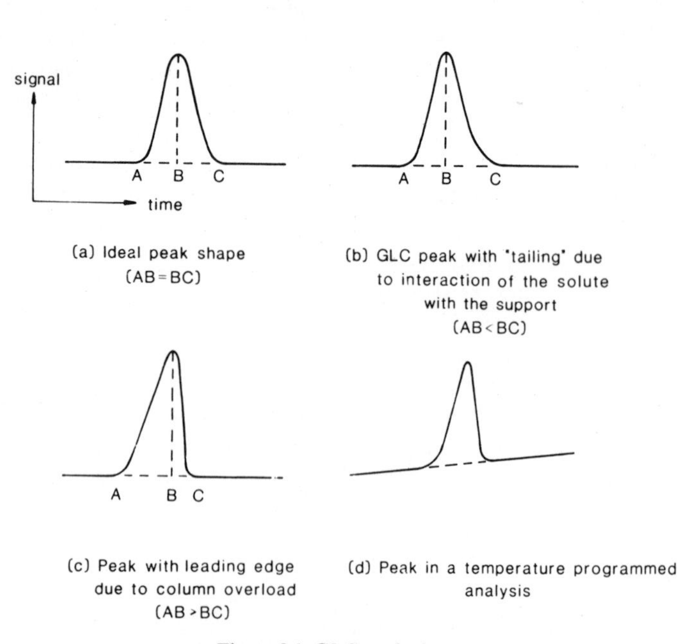

Figure 3.1 GLC peak shapes.

Diffusion causes GLC peaks to broaden into gaussian curves, although the ideal curve shape is sometimes distorted by interaction of a component with the solid support or by column overload or temperature programming, as shown in Figure 3.1.

3.3 The chromatographic system

The basic elements required for GLC are shown in Figure 3.2. Carrier gas from a cylinder of compressed gas first passes through a mass flow controller with flow rates of 20–60 ml min^{-1} being commonly used for a packed column system. Faster flow rates may be used to accelerate the analysis if the components of the mixture are well separated. Very low flow rates of 0·5–1·5 ml min^{-1} are commonly used in capillary GLC and these flow rates are often achieved using pressure control rather than flow control. The gas passes through the column, which is located in an oven, and then flows through the detector. The signal from the detector is amplified and recorded on a chart recorder or electronic integrator. The temperatures of the injector block, oven and detector block are controlled independently.

Nitrogen is the most common carrier gas for packed-column analysis, although other inert gases, including argon, carbon dioxide, helium and hydrogen, may also be used. Hydrogen is the preferred carrier gas for capillary GLC because it gives the best resolution due to its low viscosity and hence high diffusion coefficient. However, helium is an acceptable alternative and is commonly used. Nitrogen should be avoided as a carrier gas for capillary chromatography because it leads to a significant loss of resolution. This is illustrated in the analysis of fatty acid methyl esters shown in Figure 3.3, where the double bond positional isomers 20:1 $n-9$ and 20:1 $n-7$ are separated if hydrogen is used as the carrier gas, but not if nitrogen is used.

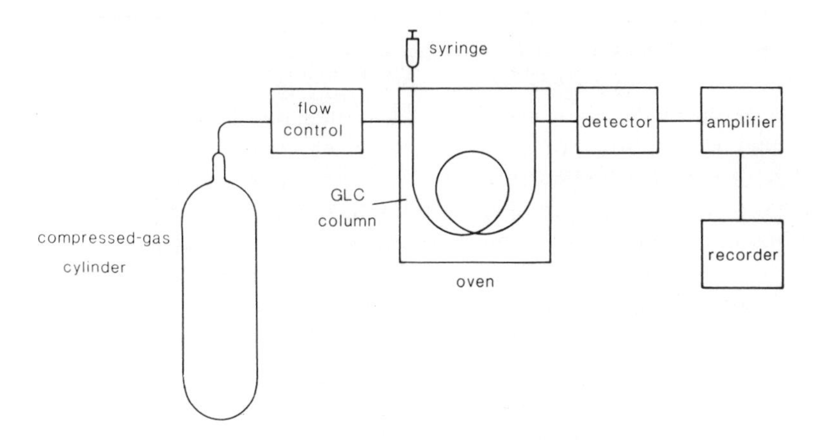

Figure 3.2 The chromatographic system.

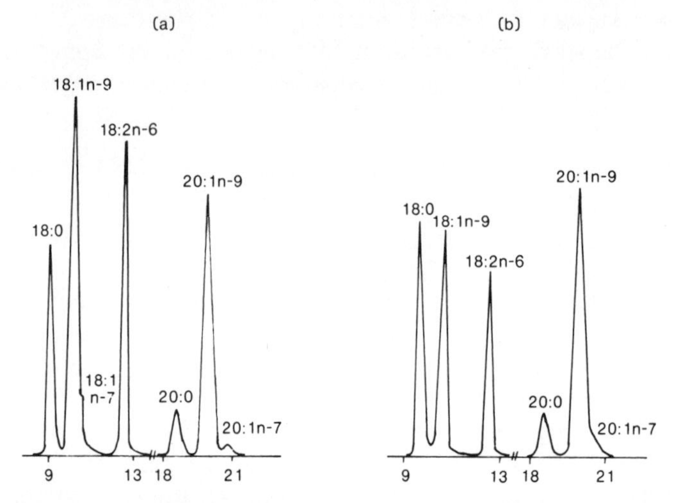

Figure 3.3 Effect of carrier gas on the chromatogram of rapeseed oil fatty acid methyl esters: (*a*), hydrogen; (*b*), nitrogen. WCOT column (46 m × 0.25 mm) with BDS as stationary phase. Inlet pressure 3.5 kg cm^{-2}. Temperature (*a*) 150 °C, (*b*) 170 °C. Redrawn with permission from Mayzaud and Ackman (1976).

Mixtures are applied to GLC columns as dilute solutions in volatile solvents. The most common sample application procedure involves the injection of about 1 μl of solution with a syringe. The sample is injected through a rubber septum, either directly onto the head of the column or into a heated space, with the carrier gas transporting the sample onto the column. Injection into a heated space generally requires the use of an inlet temperature about 50 °C higher than that of the column. It is important that the sample is still stable at this elevated temperature, and this procedure is often avoided with high boiling mixtures where decomposition may occur. Automatic samplers are now available, allowing the injection of samples when the instrument is unattended. This facility is particularly useful where a large number of routine analyses are required.

Other methods of sample application include thermal desorption of volatiles trapped on porous polymers, which is a valuable technique in headspace analysis (see Section 3.9.1). Gas samples can be injected with a gas-tight syringe, but introduction of gas from a calibrated sample loop linked to a multiport rotary valve is a more precise procedure. Biological macromolecules which are too involatile or thermally labile to be analysed directly by GC can be pyrolysed prior to analysis. This involves rapid heating of the solid sample to a temperature of 500–800 °C in a pyrolyser linked to a gas chromatograph. The fragments generated during pyrolysis can be used as a fingerprint to identify and quantify the components of the sample. Pyrolysis–GC has been useful in the analysis of synthetic polymers and micro-organisms.

The column is located in an oven with a low thermal capacity to allow rapid

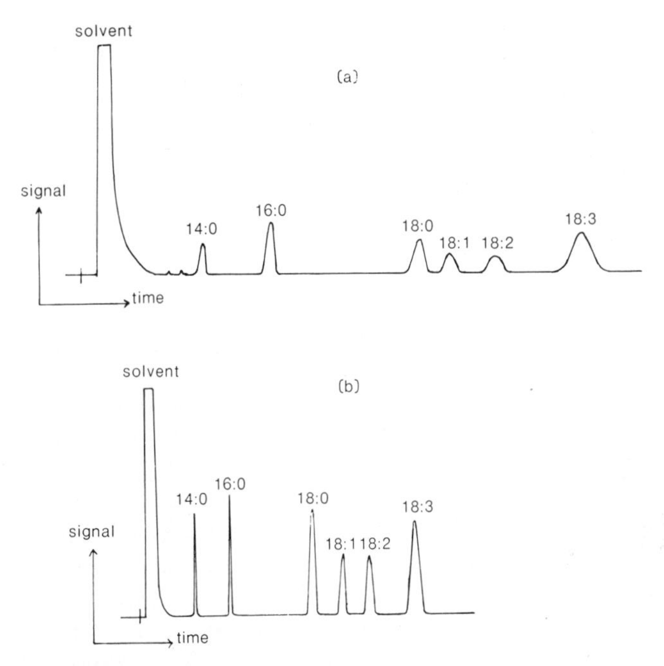

Figure 3.4 Chromatograms of a fatty acid methyl ester mixture analysed on a Carbowax 20 M packed column at (*a*) 160 °C and (*b*) 180 °C.

heating or cooling with virtually no temperature overshoot. The oven may be operated isothermally or with a temperature programme. Temperature programming is required if a mixture contains components with short retention times, as well as components with very long retention times at a given temperature. An increase in temperature shortens the retention times of the slow-moving components, but a low temperature is still required for the separation of the fast-moving components. Hence, the use of a temperature programme involving a holding period at a low temperature followed by a steady rate of temperature increase with a holding period at a higher temperature would be useful in this case. The effect of temperature on the analysis of fatty acid methyl esters is shown in the chromatograms in Figure 3.4.

3.4 GLC columns

The separation of a mixture is mainly dependent on the column and the temperature being used. Packed columns consist of a glass or metal (usually stainless steel) column with internal diameters of 2–5 mm and lengths of 0·5–8 m. Glass has the advantage that it is more inert than metal, and also the packing can be inspected visually. Metal columns have the advantage that they are not fragile, but gaps can develop in the packing during use, and these cause a loss of resolution without the cause being obvious to the analyst. In addition, metal columns can lead to the decomposition or isomerization of reactive compounds. Recently glass-lined metal tubing has been developed for use in packed-column chromatography. This tubing combines the inertness of glass with the strength and robustness of metal.

The column is packed with fine particles of an inert support coated with the stationary phase. The weight percentage of stationary phase on the support is generally in the range 1–20%. The support must be chemically inert to avoid adsorption of eluting components which leads to tailing. It must consist of uniform particles with a large surface area and must be capable of being uniformly wetted by the stationary phase. The most commonly used supports are prepared from diatomaceous earth, although other support materials including PTFE, glass beads and porous polymers are sometimes used. The inertness of a support can be improved by acid washing and by treatment with dimethyldichlorosilane. Acid washing removes mineral impurities which can cause decomposition of the sample or stationary phase, while silylation converts polar silanol groups on the surface of the support to silyl ethers which often improves peak shape, particularly for polar solutes. Column supports are sold with various particle-size distributions. Smaller particles yield columns with greater separating efficiencies, but they increase the pressure drop across the column and are not suitable for long narrow-bore columns. Supports with a particle-size distribution in the range 150–180 μm (generally described as mesh 80–100) are commonly used for packed columns. Injection of 1 μl of a solution containing solutes at about the 1% level for each component (that is, 10 μg) is suitable for packed-column analysis.

Packed columns are still widely used in applications where high resolution

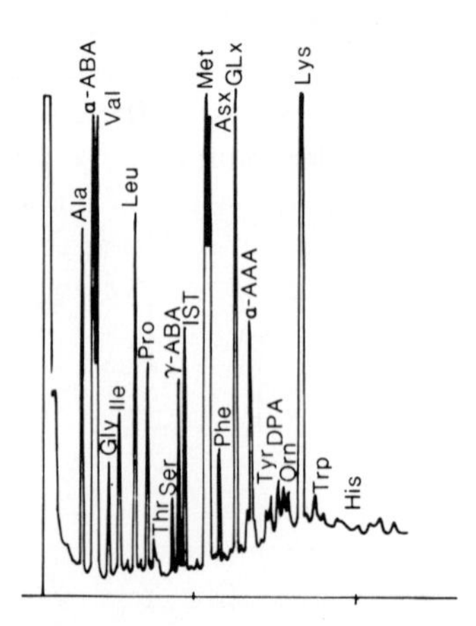

Figure 3.5 Packed column analysis of amino acids in a fermentation broth after 48 h of cultivation. Redrawn with permission from Gammerith *et al.* (1985).

is not necessary because of their relatively low cost and the fact that they can separate larger amounts of material than capillary columns. A recent report by Gamerith *et al.* (1985) describes the use of packed-column GC to monitor the formation of free amino acids in a fermentation broth during the growth of *Clostridium oncolyticum* M 55. The chromatogram shown in Figure 3.5 shows good separation of the amino acids of interest after derivatization to the N(O, S)–acyl alkyl esters.

Capillary columns are often used instead of packed columns for the analysis of complex mixtures. They give much better resolution and retention times are also generally shorter (compare Figure 3.4 and 3.6). There are two versions of capillary columns in common usage. The most common type is the wall-coated open tubular (WCOT) column, although surface-coated open tubular columns (SCOT) are also used. WCOT columns contain the stationary phase as a thin film coated onto, or chemically bonded to, the wall of a long glass or fused silica capillary. Typical capillary dimensions are 0·1–0·4 mm i.d. in a column of length 8–100 m containing a film of thickness 0·2–2 μm. SCOT columns contain the stationary phase coated onto a porous layer deposited on the inside wall; they have greater surface area per metre compared with WCOT columns and the length is generally limited to less than 25 m.

The main drawbacks of capillary columns are the high cost of the column and the smaller sample size required to avoid overloading the column. Wide-bore WCOT columns have recently been developed with typical dimensions of 0·75 mm i.d., 10–60 m length, containing a film of thickness up to 1 μm. These columns allow the analysis of more concentrated mixtures at higher carrier gas flow rates, but the resolution is inferior to that of a narrow-bore WCOT column. If the column is prepared from glass, it is extremely fragile and requires great care in manipulation. However, this limitation is not present with fused-silica columns which are very flexible. Columns in which the stationary phase is coated onto the capillary have low temperature limits and short lifetimes compared with packed columns. The columns deteriorate by oxidation of the stationary phase, by loss of the stationary phase during use due to bleeding from the end of the column, or by the accumulation of high-

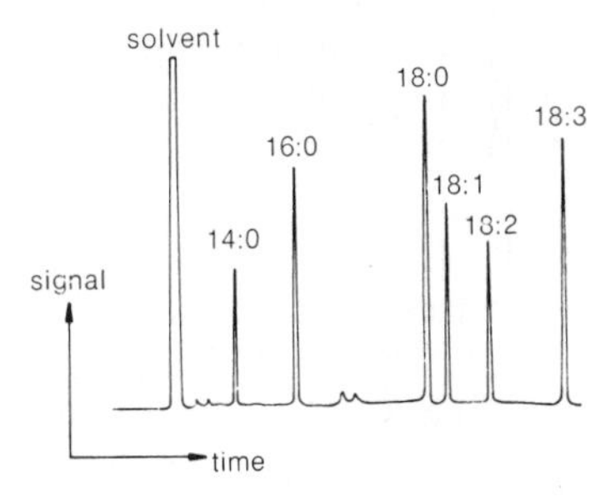

Figure 3.6 Analysis of a fatty acid methyl ester mixture on a WCOT column, stationary phase Carbowax 20 M, length 100 m.

boiling materials in the column. The first of these problems is minimized by the use of an oxygen trap in the carrier gas line to remove traces of oxygen. The development of chemically bonded WCOT columns has reduced the extent of the other two problems. The stationary phase is chemically bonded to the wall of the column, and the column can be cleaned periodically by washing with an organic solvent. However, capillary columns containing chemically bonded stationary phases of very high polarity are not available at present.

High-boiling impurities are removed from packed columns or WCOT columns containing a coated stationary phase by heating the column with carrier gas flowing, at temperatures close to the limit for the appropriate stationary phase. However, if this is performed too often with WCOT columns, it will shorten the lifetime of the column due to bleeding of the stationary phase.

Several injection techniques are used with capillary columns:

(a) *Split injection.* This injection mode was developed to allow the injection of normal sample size and concentration. A sample is injected by syringe through a septum into a heated space which is swept with carrier gas. The carrier gas is then split into two streams, one of which is vented while the other passes onto the column. The split ratio can be adjusted with a needle valve with ratios in the range 10:1 to 200:1 being commonly used. The procedure is very effective, except for the injection of mixtures which include components with a wide range of boiling points, where non-linear splitting of the high-boiling components may occur and lead to inaccuracies. Bleed of components from the septum onto the column, which may interfere with the analysis, is minimized by a split-injection technique.

(b) *Splitless injection.* In this technique, a solution is injected into a heated space followed by transport of the whole sample onto the column. This leads to considerable tailing of the solvent, which may interfere with the analysis of volatile components. In order to overcome this problem, cold trapping is commonly used, in which the column is kept at a low temperature, commonly near ambient, during the injection. The solvent is volatile and passes along the column while the components of the mixture are concentrated at the head of the column. After a short delay of about 30 s, the column is heated to the analytical temperature. It is generally helpful to vent residual solvent vapours at the end of the delay period.

An alternative splitless injection procedure uses the Grob solvent effect. Use of a column temperature about 25 °C below the boiling point of the solvent during the injection causes the solvent to condense at the start of the column and act as a second stationary phase. This concentrates the other sample components and residual solvent vapours are vented after about 20 s to avoid tailing of the solvent peak. Good resolution can be achieved by this procedure.

(c) *On-column injection.* Injectors which allow injection of sample directly onto the capillary column are now available. Syringes with metal or fused-silica needles are used, depending on the injector design. This procedure is particularly useful with mixtures containing high-boiling components, because the whole sample is applied to the column.

(d) *Thermal desorption procedures.* Trapping of volatiles onto solids followed by thermal desorption into the gas chromatograph is a valuable procedure in vapour analysis, because it allows concentration of the volatiles prior to analysis. This is discussed in more detail in Section 3.9.1.

3.5 Principles of separation

The principles of separation in GLC are similar to those described for liquid chromatography in Section 2.2. The separation of two components in a mixture is dependent on the difference in their retention times Δt_r and the mean peak width w_b of the bands. In order for base-line separation to be achieved, the resolution, $R_s \geqslant 1{\cdot}5$ (see Section 2.2). However, accurate quantification of overlapping peaks can often be achieved, and consequently a smaller separation is acceptable.

3.5.1 *Chromatographic retention*

The retention time of a component is dependent on the temperature of the analysis, the carrier-gas flow rate and the type of column being used. The temperature of the analysis has a major effect on the retention time, with an increase in temperature causing a reduction in retention time. Generally, the separation of components increases with a reduction in the analytical temperature and consequently the analytical temperature chosen represents a compromise between the high temperature required for rapid analysis and the low temperature required for good separation. Higher carrier-gas flow rates lead to a reduction in the retention time. The main column variables that affect retention of components include identity of the stationary phase (see Section 3.6), concentration of stationary phase, and the length of the column. The concentration of a stationary phase used in a packed column is conventionally expressed as a percentage coating on the support, and therefore the density of the support is also important, since the mass of stationary phase present is higher with a more dense support.

Capillary columns contain a very small mass of stationary phase per unit length of column compared with packed columns. Even with much longer columns, the retention times of components separated on a capillary column are often less than on a packed column containing the same stationary phase.

3.5.2 *Band broadening*

Several factors affect the peak width. These can best be understood by considering the van Deemter equation:

$$H = A + B\bar{u}^{-1} + C\bar{u}$$

where H is the height equivalent to a theoretical plate, $\bar{u}$ is the mean linear gas velocity and A, B and C are constant for a given column at a given temperature. The constant A is the eddy diffusion term. It describes the broadening of the peak produced by the variation in gas velocity in the porous structure of packed columns. The second term $B\bar{u}^{-1}$ represents the broadening of the peak produced by longitudinal diffusion of solute molecules in the gas phase during their passage through the column. The third term $C\bar{u}$ is related to the resistance to mass transfer in the column, which retards the equilibration of solute molecules between the gas and the stationary phase. The theoretical van Deemter curve for a capillary column is compared with the experimental curves found for the carrier gases hydrogen, helium and nitrogen in Figure 3.7.

The height H is affected by the support particle size, the column diameter and length, the carrier gas used, the temperature, the carrier-gas velocity, the

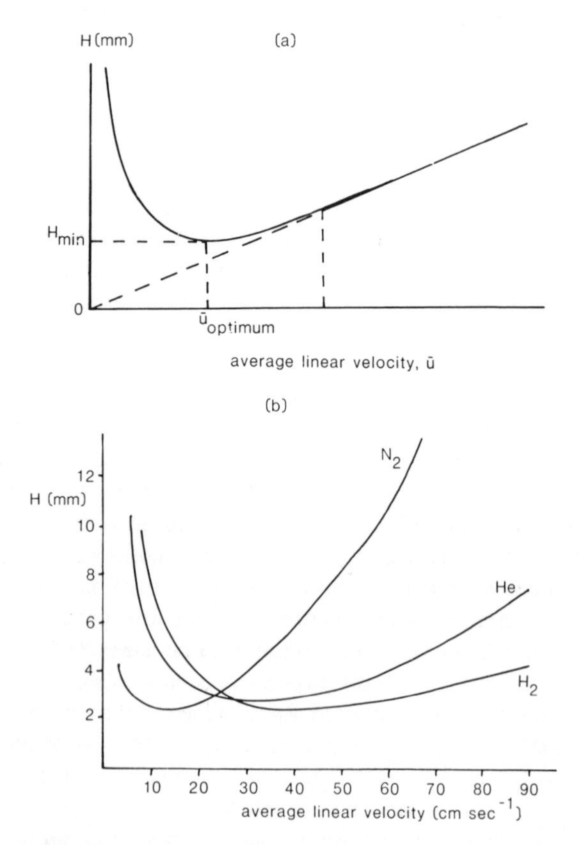

Figure 3.7 (*a*) Theoretical van Deemter curve for an open-tubular column. (*b*) Experimental van Deemter curves obtained on a 50 m × 0.25 mm glass capillary with different carrier gases. Redrawn with permission from Rooney *et al.* (1979).

sample size and the viscosity of the stationary phase (and hence the diffusion coefficient of the sample within it). A reduction in the particle diameter of the support reduces eddy diffusion and hence reduces H, causing a reduction in peak width. Thus small particle sizes are preferred, but a reduction in particle size increases the pressure drop across the column making it more difficult to achieve the optimum gas velocity. In the case of capillary columns, laminar flow occurs and hence $A = 0$.

The second term in the van Deemter equation becomes significant at low carrier-gas flow rates. The parameter B is proportional to the diffusion coefficient of the solute in the gas phase. It varies with the solute, temperature, pressure and nature of the carrier gas. The diffusion coefficient is higher in hydrogen or helium than in nitrogen or argon.

The mass-transfer term C is the dominant factor contributing to peak broadening at high flow rates. C corresponds to the sum of the mass transfer terms for the gaseous (C_g) and liquid phases (C_l).

For open tubular columns, the following expressions have been derived:

$$C_g = \frac{1 + 6k + 11k^2}{24(1 + k)^2} \frac{r^2}{D_g}$$

$$C_l = \frac{k^3}{6(1 + k)^2} \frac{r^2}{K^2 D_l}$$

where D_g and D_l are the diffusion coefficients in the gas and liquid phases, r is the internal radius of the column, K is the partition coefficient and k is the partition ratio (the ratio of the weight of solute in the stationary phase to that in the mobile phase).

The value of C, and hence the peak width, increases with the square of the column radius and inversely with the diffusion coefficients in the gas and liquid phases. Increase in thickness of the film of stationary phase also causes a decrease in resolution because k increases. The analysis of the mass-transfer term is more complex in the case of packed columns, but it can be shown that the resolution increases with a reduction in column radius, a reduction in support particle size, and a reduction in the concentration of stationary phase.

Capillary columns have much better resolution than packed columns. The resolution per metre is usually better and the low pressure drop across capillary columns allows very long columns to be used. Packed columns are generally limited to about 10 000 theoretical plates, while figures of 600 000 theoretical plates have been quoted for capillary columns. The concept of theoretical plates as a measure of column efficiency becomes less meaningful with capillary columns, because the gas hold-up time t_0 is larger relative to the retention time t_r than is the case for packed columns. Since t_0 does not contribute to the separation, theoretical plate numbers should be treated with caution, but the greater efficiency of capillary columns is evident by inspection (compare Figure 3.4 and 3.6). More useful measures of the performance of capillary columns are the effective plate number and the separation number. The effective plate number, n_{eff}, is given by:

$$n_{eff} = 16(t'/w_b)^2$$

where t' is the adjusted retention time and w_b is the peak width at the base. The separation number is the number of peak widths separating the peaks corresponding to two *n*-alkane homologues.

3.5.3 *Separation of poorly resolved peaks*

It is clear from the discussion in Sections 3.5.1 and 3.5.2 that there are many variables affecting the separation of two peaks. However, if two peaks are incompletely separated, the analyst has several main courses of action available:

(a) *Reduce the temperature.* The relative separation of two peaks with retention times t_{r_1} and t_{r_2} can be shown to be

$$\ln(t_{r_1}/t_{r_2}) = \left(\frac{\Delta H_2 - \Delta H_1}{RT}\right) + C$$

where ΔH_x is the enthalpy of solvation of component x by the stationary phase and C is a constant. Since the component with the longer retention time generally has a higher enthalpy of solvation, $\Delta H_2 - \Delta H_1$ is usually positive and therefore a reduction of the temperature T leads to an increase in the relative separation of the two peaks. The retention times of the components are increased by about 5% per °C reduction of temperature.

(b) *Reduce the carrier-gas velocity.* Packed columns are commonly operated with carrier-gas velocity well above the optimum value. This leads to a reduction in resolution, as shown in Figure 3.7.

(c) *Change the column.* If a different column must be used, the main variable to be changed is the stationary phase. As discussed in Section 3.6, a wide range of stationary phases are available and separations may be greatly improved by the use of an alternative stationary phase. If it is not feasible for another stationary phase to be used because of its effect on the separation of other components of the chromatogram, the separation can be improved by the use of a longer column or a column which is more densely packed to give better resolution. A reduction of the tailing of the peaks by the use of a more inert support may also be helpful.

3.6 Stationary phases

A wide variety of stationary phases are used in GLC columns. Desirable characteristics include low volatility up to high temperatures, good thermal stability and low viscosity. A limited amount of 'bleed' of stationary phase from the column can be tolerated, but excessive bleed due to significant volatility causes a rapidly rising baseline during temperature-programmed analysis. If excessive bleed occurs, and no alternative column can be used, the effect can be minimized by the use of a dual column procedure in which two identical columns are kept in the same oven with an identical carrier-gas flow. The difference in the signal from the detectors at the end of each column is recorded. A sample is applied to one of the columns and a chromatogram with a good baseline can be obtained. Some modern gas chromatographs have an automated bleed compensation facility with a single detector system. The technique involves electronic data manipulation.

Selection of a stationary phase depends on the chemical structures of the components in the mixture to be separated. In general, non-polar stationary phases separate mixtures of non-polar components mainly on the basis of boiling points, with dipole–induced dipole interactions being significant for polar components. Polar stationary phases have specific dipole–dipole, dipole–induced dipole and hydrogen-bonding interactions with components.

Table 3.1 McReynolds' constants and details of some common stationary phases

Phase	*Structure*	*McReynolds' constants* (ΔI)*						*Applications*
		$\Delta I(1)$	$\Delta I(2)$	$\Delta I(3)$	$\Delta I(4)$	$\Delta I(5)$	Sum ($\sum \Delta I$)	
Squalane	Hydrocarbon	0	0	0	0	0	0	Boiling-point separations
OV 1	Methylsilicone	16	55	44	65	42	222	Boiling-point separations
Dexsil-300	Carborane methyl-silicone	47	80	103	148	96	474	Boiling-point separations
OV 17	50% Phenylmethyl-silicone	119	158	162	243	202	884	Semi-polar compounds, unsaturated hydrocarbons
QF1	Trifluoropropyl-silicone	144	233	355	463	305	1500	Alkaloids and carbonyl compounds
XE60	Cyanopropyl phenyl-methylsilicone	204	381	340	493	367	1785	Polar compounds
Carbowax 20 M	Polyethylene glycol	322	536	368	572	510	2308	Polar compounds
OV 275	Cyanosilicone	629	872	763	1106	849	4219	Fatty acid methyl esters, polarizable molecules

*Reference compounds: 1 = benzene, 2 = 1-butanol, 3 = 2-pentanone, 4 = 1-nitropropane, 5 = pyridine

A useful method of comparing the retention of solutes on a stationary phase is by means of the Kovats retention indices. The Kovats retention index of a normal alkane is defined as 100 times the number of carbon atoms in the molecule. The retention index for other molecules is equal to the retention index of the hydrocarbon with the identical corrected retention time. The differences in the retention indices of a set of standards between a non-polar stationary phase and a stationary phase of interest are termed the McReynolds' constants. These constants are published and provide a useful guide for the selection of a stationary phase, for the analysis of a mixture.

McReynolds' constants for a number of stationary phases are given in Table 3.1. Comparison of the data for Carbowax 20 M and Dexsil 300 illustrates the use of the McReynolds' constants. All standards are retarded

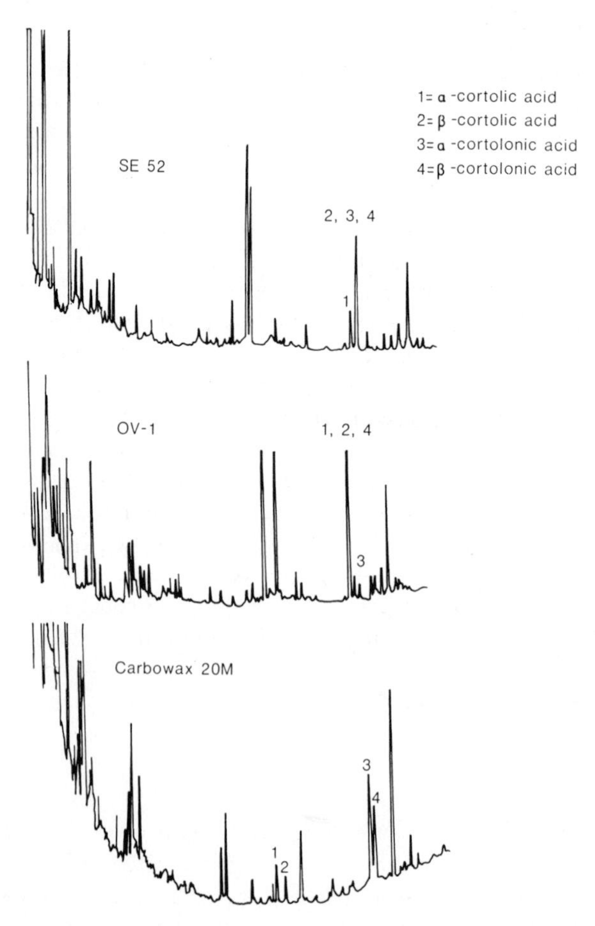

Figure 3.8 Separation of urinary cortoic acids as methylester-trimethylsilyl ethers on glass capillary columns coated with SE 52, OV-1 and Carbowax 20 M stationary phases. Redrawn with permission from Shackleton *et al.* (1980).

more strongly by the Carbowax 20 M due to increased solvation of the molecules by the polar hydroxylic stationary phase. The elution order of 1-butanol and 2-pentanone is reversed on the Carbowax 20 M phase compared with the Dexsil 300 phase, due to hydrogen bonding between the alcohol and the Carbowax 20 M.

An example of the effect of stationary phase on the separation of components in a medical study is shown in Figure 3.8. Four major acidic cortisol metabolites were analysed by Shackleton *et al.* (1980) on glass capillary columns coated with three stationary phases. The use of SE-52 or OV-1 only achieved separation into two peaks, while Carbowax 20 M separated all four components.

3.7 Gas–solid chromatography

Most gas chromatographic separations are performed by GLC, but gas–solid chromatography (GSC) is more suitable for some separations. The main problems with GSC are that solute retention times may vary with sample size, peaks are asymmetric and recovery of sample from the column may be incomplete. Also, retention times tend to be long, particularly for large polar molecules. The wide range of stationary phases available for GLC gives this technique considerably enhanced flexibility for analysis in comparison with GSC where a limited number of adsorbents may be used. The principal adsorbents used for GSC are silica, porous polymers, alumina, activated charcoal and molecular sieves. GSC does have some useful characteristics, including the fact that adsorbents are stable over a wide range of temperatures (up to 500 °C) and column bleed is virtually non-existent, so high-sensitivity detectors can be used. Also, water elutes rapidly from some packings commonly used for GSC, whereas water can lead to rapid deterioration of GLC columns due to depolymerization of the stationary phase. GSC usually gives a better separation of geometric isomers than GLC, and also hydrocarbons of low molecular mass and inorganic gases are more readily separated by this technique. GSC was used in the analysis of foreign gases in blood (Wagner *et al.* 1974). A stainless-steel column (6 ft in length) packed with Poropak-T allowed the analysis of eight contaminants including acetone, ether and methane within 8 min at 160 °C.

3.8 Detectors

The detector converts the flow of molecules passing from a GLC column into a voltage that can be monitored by a recording integrator or chart recorder. A wide variety of detectors are available. Some detectors, including the flame ionization detector, respond to the mass of eluent per unit time (mass flow), while other detectors, such as the thermal conductivity detector, respond to the vapour concentration within the detector cavity.

3.8.1 *Flame ionization detector (FID)*

The FID (see Figure 39) is the most common detector for biological samples. The effluent from the column passes into a flame consisting of hydrogen burning in air. Molecules containing oxidizable carbon (for example, $—CH_2—$groups) burn and ionization of the molecules occurs. The positively charged ions and free electrons pass between two electrodes (the flame jet and a collector) across which a potential of about 400 V is applied. The current flow across an external resistor is sensed as a voltage drop, amplified and displayed on a recorder. The jet is protected from draughts by a sleeve, and the whole detector is heated to avoid condensation of water produced by the combustion.

The response of the FID is proportional to the number of oxidizable carbon atoms. It is a mass-flow detector which is insensitive to water and many inorganic gases (such as $CO, CO_2, CS_2, H_2S, NH_3$), and it will not respond to carbonyl or carboxy groups. The detector response is proportional to sample weight over a range of 10^7. Factors affecting the response include the flow rate of carrier gas, applied voltage and the flame temperature, which is dependent on the hydrogen to air ratio. The carrier-gas flow, with a packed column, should be in the range of 20–60 ml min^{-1}. The carrier gas flow through a small-bore WCOT column is only approximately 1 ml min^{-1}, and it is usual to introduce a make-up gas such as nitrogen into the detector when WCOT columns are used, since the detector sensitivity is greater at higher flow rates.

The detection limits of the FID are approximately 5 pg s^{-1} for light hydrocarbon gases, increasing up to 10 pg s^{-1} for higher organic molecules. It is useful to convert these values into detection limits in mass units. If the analysis of a component with a retention time of 30 min, using a column of 5000 theoretical plates efficiency, is considered:

$$n = 16\left(\frac{t_r}{W_b}\right)^2$$

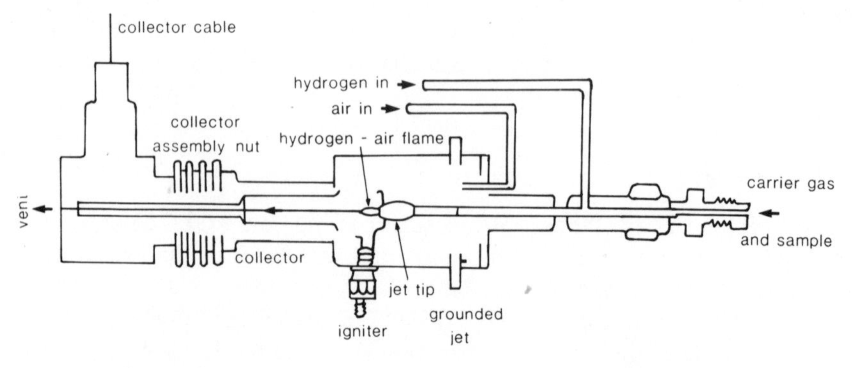

Figure 3.9 Flame ionization detector (courtesy of Perkin-Elmer Ltd).

Therefore

$$W_b = 1{\cdot}7\ \text{min}$$

The mass of material required for a detection limit of $10\,\text{pg}\,\text{s}^{-1}$ is therefore 0·5 ng, since the peak area is approximately half base × height. A realistic detection limit is usually somewhat higher than this in practice, since losses may occur due to irreversible adsorption in the chromatographic system. Also much larger masses must be analysed for the peak area to be measured and the component quantified.

If a capillary column with 100 000 theoretical plates is used for the analysis with the retention time remaining at 30 min, the theoretical detection limit becomes 0·1 ng. If a split-injection procedure is used with a capillary column using a split ratio of 50:1, the sensitivity of the analysis is less than that of a packed column by a factor of 10 times, but if a splitless procedure is used, the sensitivity is about 5 times better than that of the packed column.

3.8.2 *Thermionic detector (TD)*

The TD (Figure 3.10), also called the alkali flame ionization detector, has been developed for the analysis of compounds containing phosphorus or nitrogen with a detection limit of $10^{-13}\,\text{g}\,\text{s}^{-1}$. A bead of a non-volatile alkali metal salt such as rubidium silicate is heated to about 600–800 °C. Alkali metal atoms are formed which ionize in a fuel-poor hydrogen flame and are subjected to an electric field. The current produced is proportional to the number of ions and is enhanced by the passage of molecules containing halogen, phosphorus or nitrogen. The sensitivity to phosphorus or nitrogen compounds is generally better than 3000:1 compared with carbon.

3.8.3 *Flame photometric detector (FPD)*

The FPD (Figure 3.11) is a mass-flow detector operating on the principle of a flame photometer (see Section 11.2). The column effluent passes into a hydrogen-enriched low-temperature flame inside a shielded jet. Combustion

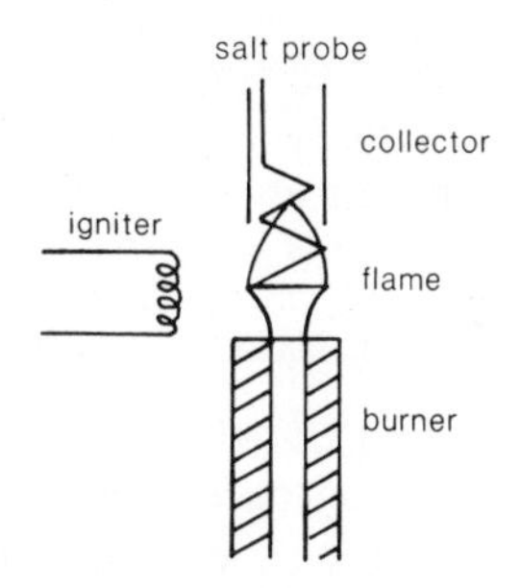

Figure 3.10 Schematic diagram of a thermionic detector.

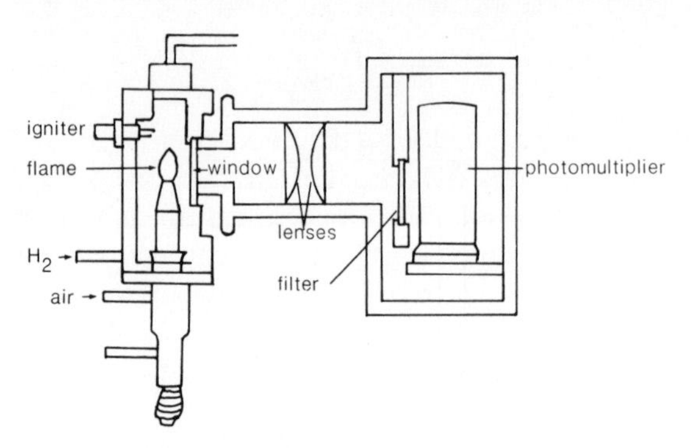

Figure 3.11 Flame photometric detector.

of molecules leads to characteristic low molecular weight species in excited states. Phosphorus-containing molecules form HPO*, while sulphur-containing molecules form S_2^*. These two species return to the ground state with the emission of radiation at 526 nm and 394 nm respectively. Passage of the emitted radiation through a narrow bandpass filter into a photomultiplier tube allows the emission at a particular wavelength to be quantified. The main applications of the FPD are in detection of compounds containing phosphorus and sulphur with detection limits of $2 \times 10^{-12}\,\mathrm{g\,s^{-1}}$ and $5 \times 10^{-11}\,\mathrm{g\,s^{-1}}$ respectively. The detector response increases linearly with sample weight over a range of 10^4 for phosphorus and over a 10^3 range on a log–log basis for sulphur. Applications of the FPD include analysis of pesticides, food volatiles and air pollutants.

3.8.4 *Electron capture detector* (*ECD*)

The ECD (Figure 3.12) consists of a chamber containing two electrodes with the carrier-gas stream passing between them. A radioisotope which emits high-energy electrons is located next to the cathode. Radioactive sources include tritium adsorbed in titanium, or ^{63}Ni as a foil or plated on the inside of the cathode chamber. The high-energy electrons produce large numbers of low-energy secondary electrons in the GC carrier gas, and these are collected by the anode. Hence a standing, or background, current occurs in the absence of eluting molecules. Electronegative species which elute from the column can capture an electron to form a negatively charged ion, and this process causes a reduction in the background current.

$$CX + e^- \rightarrow CX^- + \text{energy}$$

There are two designs for the ECD in which the anode and cathode occur as parallel plates or as concentric cylinders. The carrier gases used are nitrogen, or preferably argon containing 5–10% by volume of methane which minimizes sample ionization arising from metastable species produced from the argon.

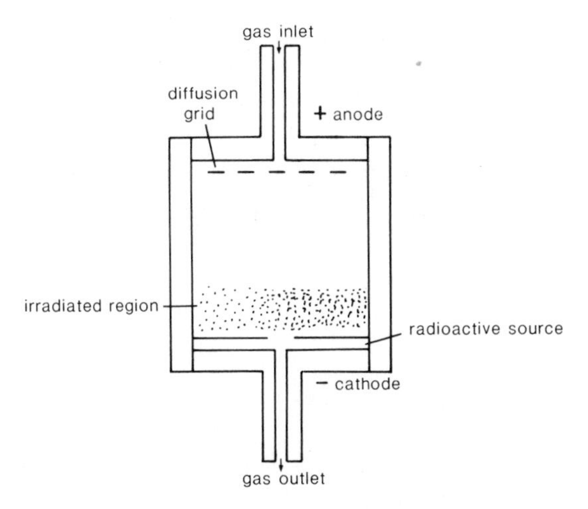

Figure 3.12 Electron capture detector.

The detector voltage of 30–50 V is usually applied as a sequence of pulses lasting 0·5–1 μs with a gap of 100–150 μs between the pulses. The use of voltage pulses which collect the mobile electrons but not the heavier anions prevents the formation of contact potentials and space-charge effects. The pulse reduces the electron concentration to zero, and the electron concentration builds up between the pulses. This mode of operation enables the free electrons to reach thermal equilibrium with the gas molecules and a good signal-to-noise ratio is achieved. The ECD is mainly applied to halogenated molecules and has proved extremely useful in the analysis of chlorinated pesticides, polychlorinated biphenyls and chlorinated hydrocarbons where concentrations of $10^{-14}\,g\,s^{-1}$ can be detected.

3.8.5 *Thermal conductivity detector (TCD)*

The TCD (Figure 3.13) consists of a cavity in a metal block with a coiled filament passing through the middle. The filament is commonly composed of platinum, nickel, tungsten or alloys of platinum or tungsten. It is heated by a direct current and the filament resistance is monitored. If the gas surrounding the filament contains an eluting compound, the thermal conductivity changes from the background value and this causes a change in the filament temperature and hence the resistance. Hydrogen or helium is used as the carrier gas, since both have a high thermal conductivity. The TCD has the advantage that it is simple and virtually universal, responding to both organic and inorganic compounds. Compounds containing metal atoms or halogens are the main classes for which the response factor deviates from that of hydrocarbons, but for most organic compounds the response factor is similar. The main drawback to the TCD is that the sensitivity is less than that of the

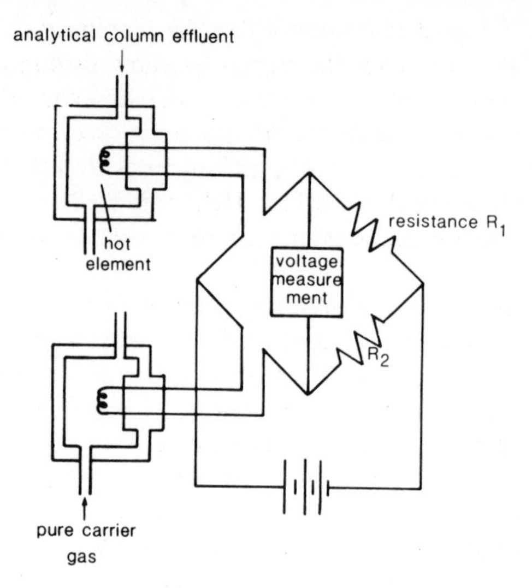

Figure 3.13 Thermal conductivity detector and associated circuitry.

other common detectors, with a limit of approximately 1 ng ml^{-1}. The TCD has applications in the analysis of gases which are not detected by the FID, such as NH_3, H_2S and CO_2.

3.8.6 *Mass spectrometer*

GC–MS represents a very powerful combination of separation and structural identification techniques. Although the system is expensive, GC–MS is required for many non-routine studies involving the analysis of complex mixtures. The advantages of a mass spectrometer are that it provides a sensitive, specific and universal method of detection. Full structural identification of unknown components is often possible from the mass spectrum. MS is similar in sensitivity to the FID or EC detectors, but is not limited in its application as these detectors are.

The most significant problem that had to be overcome in the development of GC–MS was the large difference in operating pressures of the chromatographic and detector systems. Gas chromatographs are operated with outlet pressures of 760 Torr (10^5 Pa) while mass spectrometers require pressures of $< 10^{-5}$ Torr (10^{-3} Pa). Thus interfacing the two instruments requires a 10^8-fold reduction in carrier-gas pressure, but loss of sample at the interface must be minimized. In fact, consideration of gas pressures exaggerates the problems. The gas flow rate into a mass spectrometer should be in the range 0·1–2 ml min^{-1}. This compares with flow rates in packed GC columns which are typically in the range 20–30 ml min^{-1} with higher flow rates being used for some analyses. These figures indicate that reduction in flow rate of 10–20 times is often adequate for packed-column GC–MS. Capillary columns are even more suited for GC–MS, since the carrier-gas flow through a narrow-bore capillary column is generally $< 1\,\text{ml min}^{-1}$ and therefore the outlet of the capillary column may be led directly into the mass spectrometer. Wider bore capillary columns often operate at higher flow rates, and the carrier-gas flow may need to be reduced at the interface. Helium is used as the carrier gas for GC–MS systems. Details of the interface and mass spectrometer are given in Chapter 12.

Mass spectra (usually low resolution) can be recorded for each eluting component, but selected ion monitoring (SIM) or mass fragmentography is an alternative method of analysis. A limited number of ions (usually $\leqslant 6$) are monitored and a selected ion chromatogram is produced. This procedure allows compounds containing specific ions in their mass spectrum to be selectively detected. Instruments using SIM detectors are available commercially at much lower prices than mass spectrometers. SIM also has the advantage that complex chromatograms can be simplified by this procedure. If two components co-elute in a chromatogram, the use of SIM can allow each compound to be quantified by monitoring two mass peaks simultaneously.

3.8.7 *Infrared detector*

In recent years, infrared (IR) detectors have been developed for use with both capillary and packed-column analyses. Rapid data acquisition is an essential feature of the detector, and therefore Fourier transform IR detectors are used. Details of the detection principle are given in Chapter 8. In a typical GC–IR system, the effluent from a packed column is passed down a heated transfer line into a heated IR measuring cell called the 'light pipe'. A fused-silica capillary column can feed effluent directly into the light pipe (Figure 3.14). At the light-pipe entrance, accelerating gas is added to the GC effluent to maintain the linear gas velocity through the GC–IR interface. At the end of the light pipe a

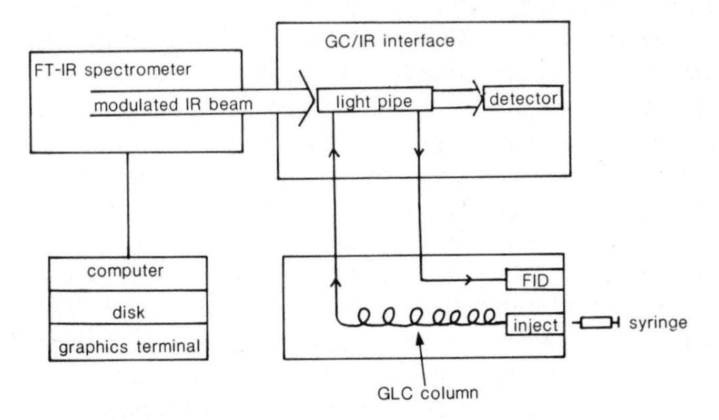

Figure 3.14 Schematic diagram of a capillary GC–FTIR system.

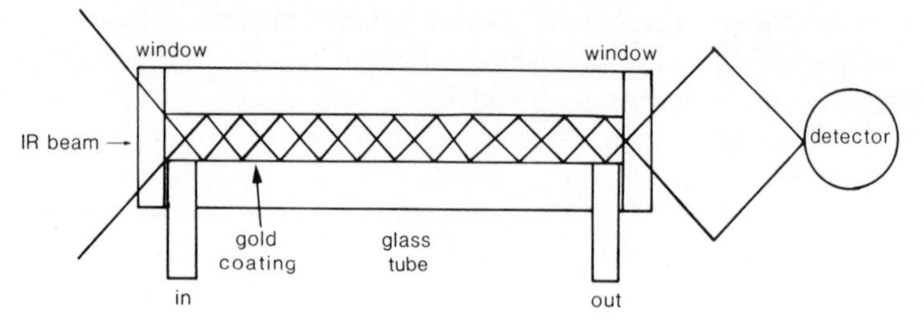

Figure 3.15 Schematic diagram of a light pipe for an FTIR detector for capillary gas chromatography.

fused-silica transfer line is used to guide part of the effluent back to the FID. The light pipe (Figure 3.15) consists of a gold-coated glass tube with IR transparent windows. The IR beam undergoes multiple reflections by the gold coating and is focused onto the element of a mercury–cadmium–tellurium detector cooled in liquid nitrogen. The volume of the light pipe for high resolution GC–FTIR work is less than 0·1 ml, although larger volumes can be used if packed-column work is also performed on the same instrument. The mass of compound detectable by FTIR is as low as 5 ng for molecules with strong IR absorption.

3.8.8 *Other detectors*

A number of other detectors are used in GLC analysis. These include the helium detector, the photoionization detector, the radioactivity detector and the Coulson conductivity detector. It is common practice for the column effluent to be split between two detectors, or for a non-destructive detector to be used in series with a second detector.

3.9 Sample preparation

The preparation of biological samples for GLC analysis generally involves some of the techniques described in Section 2.5. Sample clean-up techniques including selective extraction, TLC, column chromatography and distillation are often used.

3.9.1 *Headspace analysis*

The most widely used method of sample introduction into a gas chromatograph is injection of a dilute solution in an organic solvent with a syringe. However, GC is ideally suited to the analysis of volatile components in the environment or in the headspace above biological materials. Analysis of volatiles is important in various areas including determination of flavour and odour components in foods, alcohol or volatile metabolites in blood, and

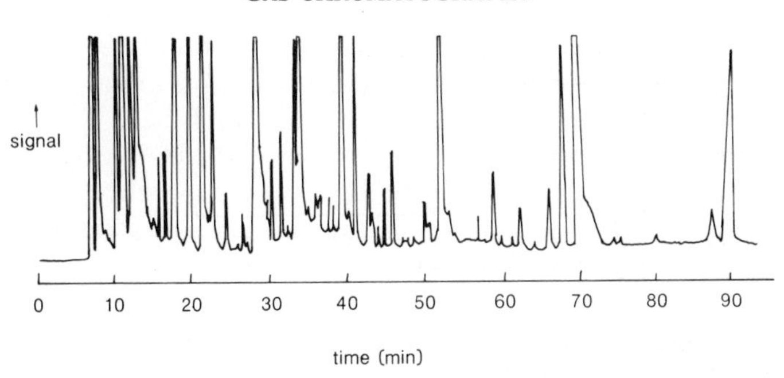

Figure 3.16 Chromatogram of extract of tawny port wine. Redrawn with permission from Williams *et al.* (1983).

volatile toxic components in the environment. Headspace analysis most simply involves manual or automatic injection of a sample of the vapour with a gas syringe. However, this technique is often insufficiently sensitive, since many volatile components may be present at very low concentrations in the headspace but may nevertheless be of importance in the biological system. Therefore, headspace-concentration techniques have been developed in which the volatile components are trapped on an adsorbent which is commonly a porous polymer, such as Porapak Q or Tenax GC, or activated charcoal. Relatively large volumes of a pure inert gas such as helium may be used to sweep the volatiles onto the trap. Improved trapping procedures involve the use of circulatory systems in which a small volume of gas is passed over the material continuously in an enclosed system. This avoids the build-up of impurities from the inert gas on the trap. For monitoring volatiles in the environment, tubes containing adsorbents can be exposed to the atmosphere for a period of time with sampling of the air either by diffusion or by pumping air through the trap. The volatiles may be removed from the trap by washing with a small volume of solvent, or more commonly, by thermal desorption, which involves heating the adsorbent in the inlet of the gas chromatograph to an elevated temperature at which the adsorbed volatiles are desorbed directly onto the analytical column.

Volatile components contribute to the odour and flavour of foods and therefore many applications of headspace analysis have been reported in this area. Williams *et al.* (1983) have used the technique to analyse the volatile flavour components of port wines. A mixture containing large numbers of components was trapped and analysed by capillary GLC (Figure 3.16). The components were identified using a mass spectrometer as the detector and the odour characteristics of the separated components were also assessed.

3.9.2 *Derivatization*

The formation of derivatives is often required in GC analysis. Derivatization may be employed to increase the volatility of involatile materials, to improve the stability of molecules that may otherwise decompose at elevated analytical temperatures or to improve the peak shape and separation of highly polar components which tail badly due to interaction with the support.

A common derivatization procedure involves the replacement of an active hydrogen of polar substituents including hydroxyl, carboxyl, amine and sulphydryl groups by a trimethylsilyl group ($—Si(CH_3)_3$). Several silylating agents are available, including trimethylchlorosilane, hexamethyldisilazane, trimethylsilylimidazole and bis (trimethylsilyl) acetamide. These reagents vary in reactivity, selectivity and sensitivity to water. The solvent used is commonly pyridine. Silylation generally proceeds rapidly to completion under mild conditions and the reaction mixture is injected directly into the chromatograph without isolation of the silylated products. Formation of trimethylsilyl ethers is the most common derivatization process for alcohols, but other procedures including formation of acetate esters or dimethylsilyl ethers are also employed. For example, aldose sugars have been analysed by GLC after formation of methyl ethers, trimethylsilyl ethers, acetyl or trifluoroacetyl esters, and poly-O-acetylaldonic nitriles.

3.10 Quantification

The quantitative determination of a component in a mixture analysed by GLC involves determination of the peak area followed by a procedure for converting the peak area to the concentration of the component in the mixture. Peak areas can be determined either with an electronic integrator or by a manual procedure. The manual procedures that have been used include:

(a) Area $\propto w_{1/2} \times h$
(b) Area $\propto t_r \times h'$
(c) Area $\propto w_b \times h'$
(d) Area $\propto$ mass of peak after cutting out the chart recorder trace

where h' is the height of the triangle formed by extrapolating the peak sides of maximum slope and the baseline w_b is the baseline width of this triangle, $w_{1/2}$ is the width at half-height and h is the peak height (see Figure 3.17).

The conversion of peak area to concentration can be performed either with

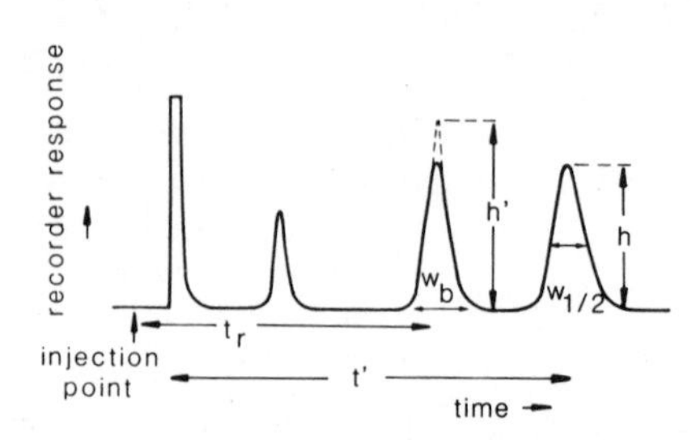

Figure 3.17 Schematic diagram of a gas chromatographic recorder trace.

an internal or an external standard (see Section 11.2.2) or by a normalization procedure. The use of an internal standard is often preferable where analysis involves manual injection of a sample by a syringe, since it is difficult to inject a precise volume. However, calibration curves prepared with external standards are often used in headspace analysis. It is essential that the effect of sample concentration on detector response is known for the concentration range of interest. Restriction of the sample concentration to the range where the detector response is linear is preferred. The area normalization procedure is widely used in GC analysis. In this method, the area of each component is reported as a percentage of the total area of all peaks:

$$\text{Area \% component X} = \frac{\text{Area peak X} \times 100}{\text{Total area for all peaks}}$$

The results reflect the relative concentrations of each component in a multicomponent mixture if all components are eluted and the components generate the same detector response per unit mass. In many cases, the detector response will not be identical for components of different structure and it is necessary to determine a response factor for each component. Relative response factors for components in a mixture can be determined by injecting a mixture containing a known mass of each component. The response factor for each component is equal to its mass percentage divided by the area percentage. For accurate work, response factors should be determined over a range of component concentrations.

Several methods of sample preparation and quantification have been employed for the determination of alcohol in blood (Jain and Cravey 1972). The procedures employed include extraction into organic solvents and quantification with a calibration curve; extraction into an organic solvent containing an internal standard; distillation, trapping and use of a calibration curve; headspace analysis and quantification with an external or internal standard; and extraction and quantification with an internal standard procedure based on peak heights. The use of peak heights for quantification is only valid if the retention times of the components remain constant. Variations in retention times may occur over a period of time, due to variations in column temperature, carrier-gas flow rates or column properties, and therefore procedures based on area measurements are preferable. The most accurate procedures are those involving headspace analysis with an internal standard.

References

Gamerith, G., Zuder, G.F.X., Giuliani, A. and Brantner, H. (1985) *J. Chromatogr.* **328**, 241.
Jain, N.C. and Cravey, R.H. (1972) *J. Chromatogr. Sci.* **10**, 263.
James, A.T. and Martin, A.J.P. (1952) *Biochem. J.* **50**, 679.
Mayzaud, P. and Ackman, R.G. (1976) *Chromatographia* **9** (7), 321.
Rooney, T.A., Aetmayer, L.H., Freeman, R.R. and Zerenner, E.H. (1979) *Am. Lab.* **11** (2), 81.
Shackleton, C.H.L., Roitman, E., Monder, C. and Bradlow, H.L. (1980) *Steroids* **36**, 289.

Wagner, P.D., Naumann, P.F. and Laravuso, R.B. (1974) *J. Appl. Physiol.* **36**, 600.
Williams, A.A., Lewis, M.J. and May, H.V. (1983) *J. Sci. Fd. Agric.* **34**, 311.

Further reading

Dickes, G.J. and Nicholas, P.V. (1976) *Gas Chromatography in Food Analysis*, Butterworths, London.
Grob, R.L. (1977) *Modern Practice of Gas Chromatography*, Wiley-Interscience, New York.
Grob, R.L. (1983) *Chromatographic Analysis of the Environment*, 2nd edn, Marcel Dekker, New York.
Jaeger, H. (1985) *Glass Capillary Chromatography in Clinical Medicine and Pharmacology*, Marcel Dekker, New York.
Jennings, W. (1980) *Gas Chromatography with Glass Capillary Columns*, 2nd edn, Academic Press, New York.
Perry, J.A. (1981) *Introduction to Analytical Gas Chromatography*, Marcel Dekker, New York.

4 Electrophoresis

4.1 Introduction

Electrophoresis is concerned with the migration of charged species within a fluid medium under the influence of an electric field. The charged components may be molecular, for example, proteins, or cells or even charged particles, but here we shall restrict our discussion to charged molecules (ions). Similarly, the fluid medium may be liquid or gaseous, but in biological systems it is only electrophoresis in the former that is of interest. It should be noted that it is mobility under the influence of an electric field that is studied and not electrode reactions as in electrochemical techniques or indeed the interaction with oppositely charged ions as in ion exchange.

The velocity of a charged molecule (or particle) in an electric field is governed by the accelerating force (the electric field) and a combination of retarding forces. Thus, the charged molecule will reach and maintain a terminal velocity in an analogous manner to an article falling under the influence of gravity. The force F exerted on the molecule is given by $F = QE$, where Q is the molecular charge and E the field strength (potential difference between electrodes divided by distance between them). The major retarding force F_s is due to friction and can be quantified by Stoke's equation:

$$F_s = 6\pi r \eta v$$

where r is the molecule (particle) radius, η the viscosity of the fluid and v the molecule velocity.

Other secondary retarding forces are also present, which result from the presence of other ions (counterions and buffer ions) in solution, whose distribution around the charged molecule of interest is distorted as a result of differences in their mobilities. Ignoring these secondary effects, the final velocity of the charged molecule will be reached when the accelerating and retarding forces are equal, that is,

$$F = F_s$$

or

$$QE = 6\pi r \eta v$$

$$v = \frac{QE}{6\pi r \eta}$$

Molecules with different charges, or sizes, will migrate at different velocities and this is the basis of electrophoretic separations. The situation may be complicated further if there is an interaction between the charged molecules and the support material. The latter is used to reduce other effects such as convection, which would disrupt the migration of the components as discrete bands. These effects will be discussed for the various techniques in Section 4.3.

4.2 Effect of pH on charge

The effect of pH on the surface charge of a molecule, such as a protein, depends on the number and type of ionizable groups present. The effect of these acidic and basic groups is that the protein will show a net zero charge at a specific pH (isoelectric point pI). Below this pH, the protein will have a net positive charge which will increase with decreasing pH, as will its mobility on electrophoresis. Above the pI, the protein will have a net negative charge and its mobility will increase with increasing pH but this time in the opposite direction. The effect of pH on molecular charge is equally important in electrophoresis as in ion exchange, as previously noted in Section 2.3.3. The same phenomenon will be observed in acids and bases, except there will be no change in sign with pH but only a change in the magnitude of the charge. For example, at a pH significantly below its pI (at least 2 pH units below), an acid will be present essentially in the un-ionized form and therefore will not take part in electrophoresis. Unfortunately, in practice, the correlation between electrophoretic mobility and charge is found to be more complex but the concept of pI values provides a useful guide.

4.3 Techniques of electrophoresis

Electrophoresis has provided the basis for a very wide range of techniques, both analytical and preparative. In many of these, differences in electrophoretic mobility are combined with other separative mechanisms to provide 'hybrid techniques'.

4.3.1 *Moving-boundary electrophoresis*

Moving-boundary electrophoresis employs a U-tube filled with buffer solution and two electrodes placed at the ends of the tube to provide the required potential gradient. The sample is introduced into the bottom of the U-tube and electrophoresis is carried out at low temperature (4 °C) to reduce problems due to convection. The separated components are located by detecting minor changes in the optical properties of the buffer solution, for example by interference effects. Moving-boundary electrophoresis provides limited resolution and now finds only restricted applications, for example in preparative separations of small amounts of proteins.

4.3.2 *Paper electrophoresis*

Paper was one of the first support materials to be used in electrophoresis and it has been used in a number of different configurations. In all of these, the principle of operation is the same, that is, the sample is applied as a spot or band onto a piece of buffer-soaked paper whose ends are connected electrically via suitable troughs of additional buffer to the electrodes. Migration and separation then takes place under the influence of the applied field. The four most commonly encountered configurations are shown in Figure 4.1. In the 'sandwich type', the support paper is held between two glass plates. This provides a limited amount of conduction to dissipate the heat generated and also reduces evaporation of the aqueous buffer. Care must be taken to avoid uneven pressure between the plates, or uneven migration will occur due to differing amounts of buffer held on the paper. In the second of these techniques, the paper, often as strips, is supported by a central pole. The only cooling is by convection and air conduction and this can lead to heterogeneity across the paper. Evaporation is reduced by placing the apparatus in a closed vessel, which will quickly saturate with water vapour. The same problems of cooling apply to the horizontal technique, but problems of movement of the buffer are reduced by the horizontal configuration. In the final technique, the paper is immersed in a non-conducting immiscible solvent, such as chlorobenzene or liquid paraffin, and great care must be taken if flammable solvents such as the latter are used. Here evaporation of the buffer is completely avoided (unless it boils!) and the immiscible solvent acts as an efficient heat sink.

The electric fields applied in paper electrophoresis are usually quite modest, of the order of 5–10 V cm^{-1}, so that a potential of 100–200 V would be required across a 20 cm strip of paper. However, for certain applications, it is advantageous to apply much higher voltages to larger pieces of paper to yield stronger electric fields. Under these conditions, heating effects become more significant and so a mechanism for cooling is essential. Thus, of the techniques

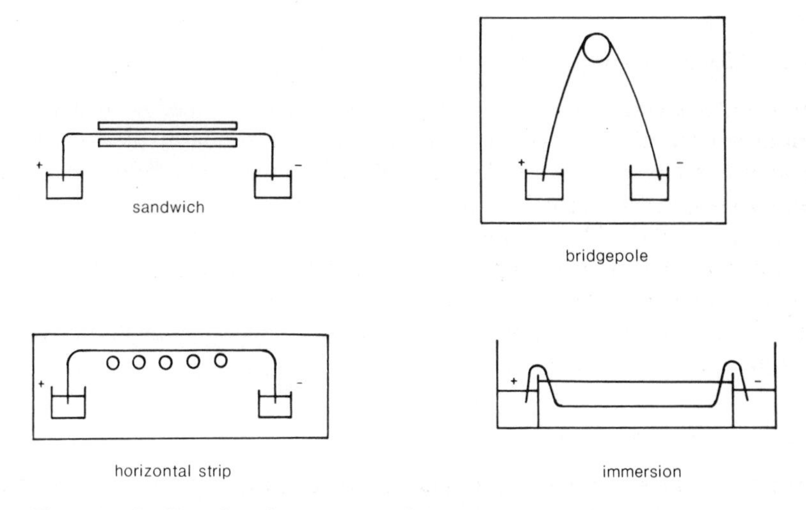

Figure 4.1 Configurations for paper (or cellulose acetate membrane) electrophoresis.

shown in Figure 4.1, only the sandwich and immersion configurations can be used. In the former, the glass plates may be replaced with large water-cooled steel plates covered with an insulating membrane to aid the removal of excess heat. High-voltage techniques have been widely used for the mapping of amino acids and peptides from protein hydrolysates. Two-dimensional paper electrophoresis is also possible, using different pH buffers for each dimension, or even paper chromatography in one dimension followed by paper electrophoresis in the second. The components can be visualized by using conventional spray reagents, for example ninhydrin for amino acids and peptides.

4.3.3 *Cellulose-acetate electrophoresis*

Cellulose-acetate membranes have now largely replaced paper as the support material for thin-layer electrophoretic techniques. The increased homogeneity of the membranes means that improved resolution can be obtained and furthermore this improvement can be achieved on a smaller scale with an increase in sensitivity. The membranes can be made transparent after electrophoresis by soaking in aqueous acetic acid prior to staining. This facilitates quantification by densitometry. However, care must be taken to ensure that the components are not washed from the membrane and some form of preliminary fixing, such as denaturation of proteins, may be required.

Cellulose-acetate membranes are not very strong, being brittle when dry and weak when wet, and this has led to the use of other thin-layer techniques where the electrophoretic support is held on an inert plate, in an analogous manner to TLC. Cellulose on glass or plastic has been used in a horizontal configuration with paper wicks making contact with the electrode troughs.

4.3.4 *Rod-gel electrophoresis*

The techniques of thin-layer electrophoresis described in Sections 4.3.2 and 4.3.3 have now been replaced, to a considerable extent, by gel techniques in which the support material is a rod-shaped gel in a vertical glass or plastic column. The sample is applied to the top of the gel and an electric field is applied via two buffer reservoirs, as shown in Figure 4.2. A wide range of

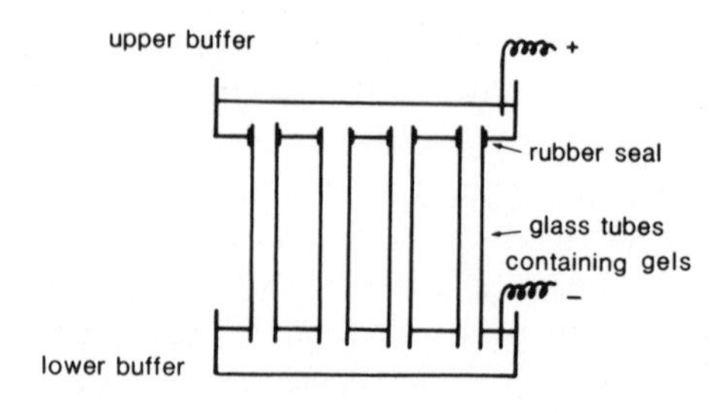

Figure 4.2 Apparatus for rod electrophoresis.

commercial apparatus is available, but the general configuration adopted is the same. The ideal gel material would be chemically stable and sufficiently viscous to eliminate convection currents, but would not impede the movement of the sample components by an interaction with the matrix. In practice, it is found that the gels used are not inert and, indeed, in many techniques, the chromatographic nature of the gel is used to enhance the separative power of the system.

4.3.4.1 *Starch gel electrophoresis.* Starch gels are prepared by heating partially hydrolysed starch (potato) in water to 95 °C and then allowing the mixture to gel in the tubes used for electrophoresis. The properties of the gel, in particular its pore-size distribution, depend on a number of factors including the starch concentration and degree of partial hydrolysis. In practice, it is found that the optimum resolution is obtained using gels containing 13–16% starch, and a suitable degree of hydrolysis may be achieved by suspending the starch in acetone containing 1% (V/V) concentrated hydrochloric acid. Commercial preparations of partially hydrolysed starch are now available. Starch gels can be used with a wide range of buffer pH's (2–11), the best results often being obtained with relatively low ionic strengths, for example 0.02 M. After electrophoresis, the gels are removed from the tubes for staining, in a similar manner to cellulose-acetate membranes. The proteins must be fixed, for example, by denaturation with 20% sulphosalicylic acid, prior to staining, or a denaturing solvent used in conjunction with the stain. A typical stain solution in the latter category would be 0.25% Coomassie Blue in 10% (V/V) acetic acid in aqueous methanol (1:1, V/V). After staining for up to 10 h, the excess stain is removed by soaking the gel in the same solvent mixture, without the dye. Alternatively, the excess dye can be removed by electrophoretic destaining. The gel is replaced in the tubes and electrophoresed using a more concentrated buffer. The proteins will not move, as they have been rendered insoluble, but the dye (charged) will migrate rapidly out of the gel.

Starch gels were widely used in the separation and identification of proteins, such as plasma proteins, but this medium has now largely been replaced by polyacrylamide, which is more reproducible in its characteristics and provides more homogeneous gels.

4.3.4.2 *Agarose-gel electrophoresis.* Agarose is a natural polysaccharide consisting of galactose and 3, 6-anhydrogalactose units derived from the agar of *Gelidium amansii.* Gels may be prepared in a similar way to starch gels by dispersing the dry polymer in boiling water and cooling to room temperature. The resulting gel has a much more porous structure than is possible with starch gels (or even polyacrylamide gels, see below) yet is still adequately rigid. Electrophoresis in agarose gels can then be considered closer to 'true electrophoresis' where the charged molecules are moving under the influence of an electric field but are not retarded by the support material. Agarose gels have been widely used for the electrophoresis of the largest macromolecules and even macromolecular complexes such as viruses, enzyme complexes, lipoproteins and nucleic acids.

4.3.4.3 *Polyacrylamide-gel (zone) electrophoresis.* Polyacrylamide gels can be prepared by the polymerization of monomeric acrylamide in the presence of N, N′-methylene-*bis*-acrylamide (*bis*-acrylamide), the latter acting as a cross-linking reagent. The polymerization is initiated by sulphate free radicals which form when ammonium persulphate is dissolved in water:

$$S_2O_8^{2-} \rightarrow 2SO_4^{-\cdot}$$

An alternative method of initiation involves free radicals formed by the UV photodecomposition of riboflavin which is added at very low levels to the acrylamide (1 mg per 100 g). Catalysts, such as tetramethylethylenediamine may also be used to increase the rate of gel formation.

The pore-size distribution of the resulting gel depends on the total amount of acrylamide used and the proportion of this which comes from the cross-linking reagent. In practice, it has been found that the minimum pore size is obtained with 5% *bis*-acrylamide, with a total acrylamide concentration in the range 8–12%. In many instances, slightly less *bis*-acrylamide is used (2–3%), providing a more open gel structure.

The mobility of a charged molecule through the gel will then be a combination of its electrophoretic mobility, which is governed by the equations in Section 4.1, and its ability to pass through the porous support. If the charged molecules are all larger than the pores available, no movement will take place, irrespective of the strength of the applied field. It is this additional separation of molecules on the basis of their molecular size that enhances the resolution of polyacrylamide-gel electrophoresis. The same effect is operative in starch gels, but in that case it is more difficult to control the pore-size distribution and hence degree of 'molecular sieving'. The apparatus used for the electrophoresis is the same as for starch gels, shown in Figure 4.2. The methods of fixing and staining are also similar, but in this case the polyacrylamide gel is transparent and so transmission densitometry is easily carried out, in an analogous manner to reflectance densitometry described for TLC in Section 2.4.2.

4.3.4.4 *Disc electrophoresis.* Disc electrophoresis, or discontinuous electrophoresis, is a development of the zone technique already described. The name is derived from the discontinuous nature of the buffer systems employed or, at least in the original method of Ornstein (1964) and Davis (1964), the discontinuous nature of the gel. In this method, the sample is applied as a gel to the top of a stacking gel, which in turn is situated on top of the separation (or running) gel. The latter gel is prepared in the same way as for zone polyacrylamide-gel electrophoresis with the same acrylamide concentration (approximately 10%) and, for protein separations, is often used with a Tris (tris(hydroxymethyl)aminomethane) buffer, pH approximately 8·5. After the running gel has set, the stacking gel (1 cm) is formed on top. This gel consists of a much lower acrylamide concentration (2–3%) and hence possesses a much wider pore structure. A further portion of the stacking-gel solution, containing the sample, is then poured in to form the sample gel. The arrangement of these gels together with typical buffer solutions is shown in Figure 4.3.

As soon as the current is turned on, a complex set of ionic migrations ensue. Glycine in the upper reservoir will be present predominantly in its anionic form and will therefore migrate towards the anode, as will the charged protein and chloride ions. Now as the glycine anions enter the stacking gel, at lower pH, they will revert to a considerable extent to their zwitterionic form with zero net charge and so will migrate only very slowly. This will lead to a reduction in the number of mobile ions in the sample and stacking gel and so the current will fall. This reduction in conductivity will lead to a high potential gradient between the leading chloride ions and the trailing glycine. The protein anions will then migrate rapidly under these conditions to form a narrow disc just behind the chloride ions. There is no tendency for the proteins to be impeded, as the stacking gel has a wide pore structure, nor for them to overtake the chloride ions as there is no ion deficiency near the chloride ions and so the electric field is diminished. The proteins then enter the running gel, where they are retarded by the finer pore structure, and so the glycine anions can 'catch up'. As the latter enter the running gel they become fully charged again and the ion deficiency, and hence local high potential gradient, is removed. The proteins are then subjected to a constant field strength in an analogous manner to the zone electrophoresis previously described, the major advantage being in improved resolution as a result of the pre-concentration.

The use of several types of gel, sample, stacking and running, clearly

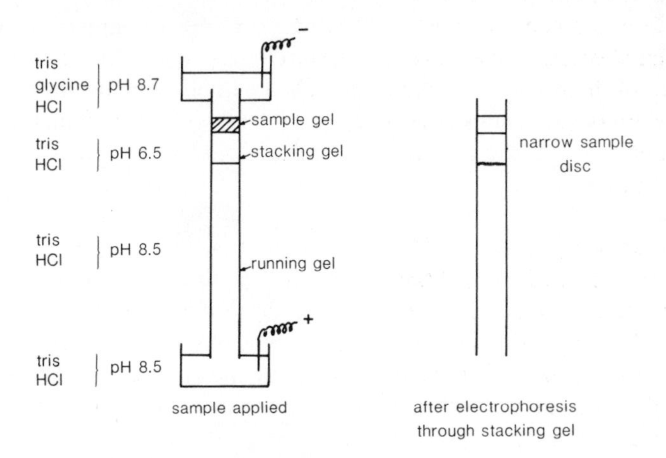

Figure 4.3 Effect of stacking gel on sample preconcentration.

complicates the technique and numerous further modifications have been successfully used based on a single-gel system. The discontinuous nature of the buffer system can be maintained with an increase in buffer concentration within the running gel, as compared with the electrode (and sample) buffer. The sample gel may also be dispensed with by applying the sample in a viscous medium of sucrose or glycerol, which helps to prevent diffusion at the top of the column prior to electrophoresis. An example of the resolution available using such a simplified disc technique is shown in Figure 4.4. for the separation of whey proteins in cow's milk.

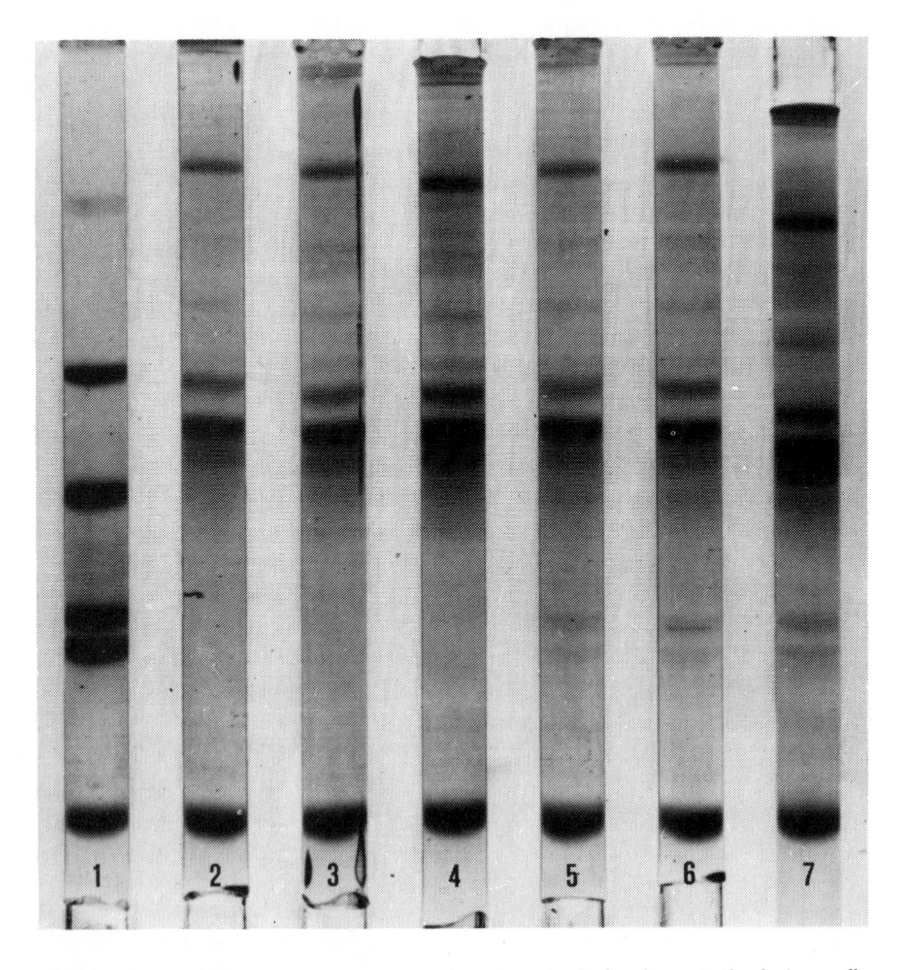

Figure 4.4 Disc electrophoresis. Sample 1, bovine whey protein fraction; samples 2–4, camel's milk; samples, 5–7, camel's milk containing bovine milk. Gel, 9% acrylamide in pH 8.9 buffer (46 g tris and 4 ml conc. HCl in 1l). Electrode buffer pH 8.3 (1.2 g tris and 5.8 g giycine in 1l). Gels stained with 0.25% Page Blue G 90 in 10% trichloroacetic acid in methanol/water (1:1). Reproduced with permission from Wilbey and Khosrowshahi-Asl, unpublished results.

4.3.4.5 *SDS polyacrylamide-gel electrophoresis.* The mobility of charged molecules in gels is a function of both their size and charge, so it is quite possible for molecules with the same electrophoretic mobility, under specified conditions, to be quite different in terms of relative molecular mass. The situation is further complicated by the fact that proteins may associate in buffer solutions and hence migrate during electrophoresis as aggregates, thereby not reflecting their true charge or relative molecular mass. One method that is commonly used to overcome these problems is to include an anionic detergent, such as sodium dodecyl sulphate (SDS), into the buffer system. SDS binds to hydrophobic sites within the protein, hence reducing the possibility of hydrophobic bonding between protein molecules. SDS also imparts a large negative charge to the protein units, the contribution from the SDS largely swamping any effect of charged groups within the protein. Under these conditions it has been found that the electrophoretic mobility of a protein is inversely related to the log of its relative molecular mass. Clearly, the exact correlation depends on the nature of the gel used and so, if the method is to be used for the estimation of relative molecular mass, the system must be calibrated with proteins of known relative molecular masses.

Protein subunits may also be bound together by disulphide bridges (cystine) and SDS alone will not cleave these. However, if the protein sample is reduced prior to electrophoresis, for example with mercaptoethanol, the subunits will be freed. Recombination of the subunits by oxidation of the thiol groups to disulphide bonds is prevented by maintaining a reducing environment in the buffers. SDS electrophoresis under these conditions then provides a very important technique for the determination of subunit relative molecular masses. An alternative approach to preventing oxidation of the thiol groups involves alkylation, for example with iodoacetic acid, and this is especially useful if the subunits are to be isolated, possibly for amino-acid analysis.

4.3.4.6 *Isoelectric focusing* (*IEF*). With the electrophoretic techniques described above, the charged molecules will continue to migrate as long as the electric field is applied, and therefore must be stopped before the components enter the buffer reservoirs. An alternative technique, isoelectric focusing (IEF), involves the migration of the charged components (proteins) to specific positions within the support material, where they are focused. A pH gradient is established across the support material between the anode and cathode by electrolysis of a suitable mixture of ampholytes. The most commonly used materials are polyamino-polycarboxylic acids with relative molecular masses of 300–600. As a consequence of the electrolysis, the anode will become the most acidic and the cathode basic, with a pH gradient between them. When the sample is introduced into the system, the proteins will migrate to a position in the gradient where they have no net charge, that is, where pH equals pI. If the proteins should diffuse away from this position, they will become charged and therefore migrate either towards the anode or cathode to return to the position where they have zero charge; in other words they will focus into a very narrow band. The resolving power of the system, in terms of the minimum difference in pI between proteins required for them to be separated, depends on the extremes of the pH gradient. It is possible to use wide-range gradients (low resolution) for preliminary analyses and then to use narrow pH ranges for more precise work. The traditional support material for IEF, which is necessary to avoid convection currents, is concentrated sucrose solution. However, this has now largely been replaced by polyacrylamide or agarose gels. After focusing, the gels are stained and quantified as with the other gel techniques. Originally IEF was carried out with short columns of gels (rods) but, as with most of the other electrophoresis techniques, slab or thin-layer configurations are becoming more widely used.

4.3.5 *Slab-gel electrophoresis*

Rod electrophoresis techniques can be advantageously translated to slab techniques in which a number of samples, and standards, may be studied simultaneously on a single piece of support material. The major advantage is in reproducibility of the electrophoretic migration, which helps identification, and quantification, as all the samples and standards will have been subjected to exactly the same environment. The most common gel material is probably polyacrylamide, although agarose is also widely used. The slabs may be cast in any convenient size, although gels larger than 10 × 10 cm are rarely required. The slabs are usually 1–3 mm thick, but there are significant advantages in using ultra-thin gels (0·1–0·25 mm). The removal of heat generated during electrophoresis is improved and also the sample is applied in

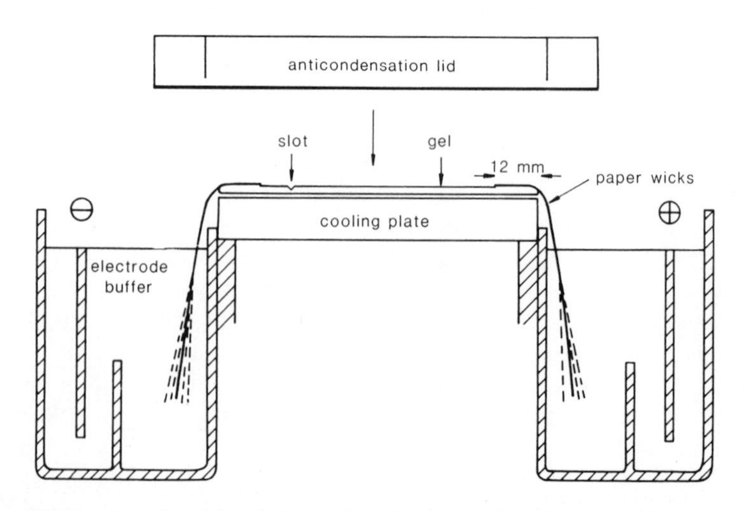

Figure 4.5 Apparatus for slab gel electrophoresis. Reproduced with permission from LKB Instruments.

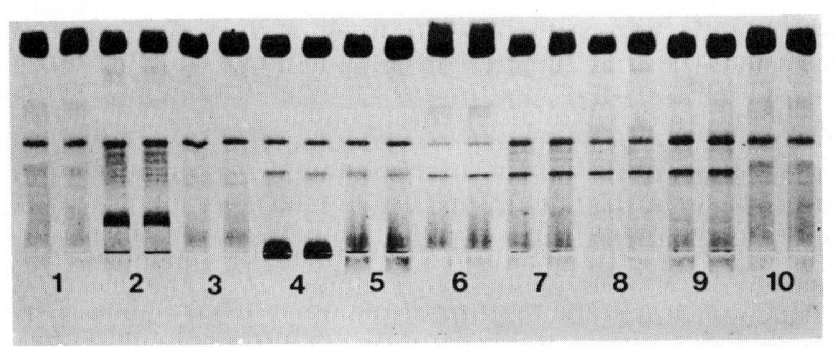

Figure 4.6 Screening electrophoresis of different human sera in a gel with 3.5% acrylamide and 5 μl slots. The samples were diluted 1:3 with 0.01 M tris-glycine. Electrophoresis was perfomed at 15 V cm^{-1}, 45 mA for 1.8 h. The samples were run in duplicate. 1, 10: Normal human sera; 2: serum from a patient with influenza; 3: serum from a patient with haemolysis; 4: serum from a patient with myeloma; 5: serum from a patient with increased lgG; 6: serum from a patient with double banded albumin; 7,8,9: unclassified pathological sera. Reproduced with permission from LKB Instruments.

a very shallow (0·1 mm or less) trough cast in the gel, which results in a more uniform sample zone. Electrophoresis may be carried out with the gel held in a vertical or horizontal position; a typical arrangement for the latter is shown in Figure 4.5. Some workers prefer a vertical orientation of the gel, as this allows large volumes of sample to be applied and is also more convenient for gradient gels.

Conventional electrophoresis, SDS electrophoresis and IEF can all readily be carried out with slab gels. Some examples of these techniques are shown in Figures 4.6–4.8 which illustrate the very high level of resolution which is possible with modern electrophoretic techniques. The SDS electrophoresis in

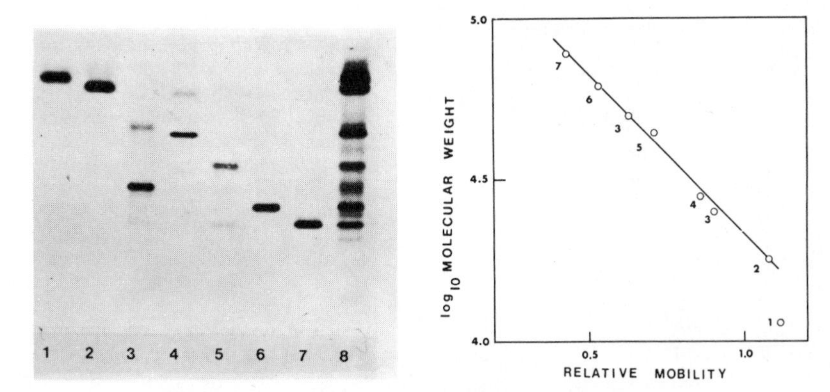

Figure 4.7 Gel: 5% acrylamide, 0.1 M phosphate buffer, 0.1% SDS, 10 μl slots. Cooling water, 5°C; 5 V cm^{-1}; 200 mA; 4 h. Sample concentration: 0.3 mg protein ml^{-1} of each component. Samples: 1, cytochrome c; 2, myoglobin; 3, γ-globulin; 4, carbonic anhydrase; 5, ovalbumin; 6, albumin; 7, transferrin; 8, mixture of proteins 1-7. Reproduced with permission from LKB Instruments.

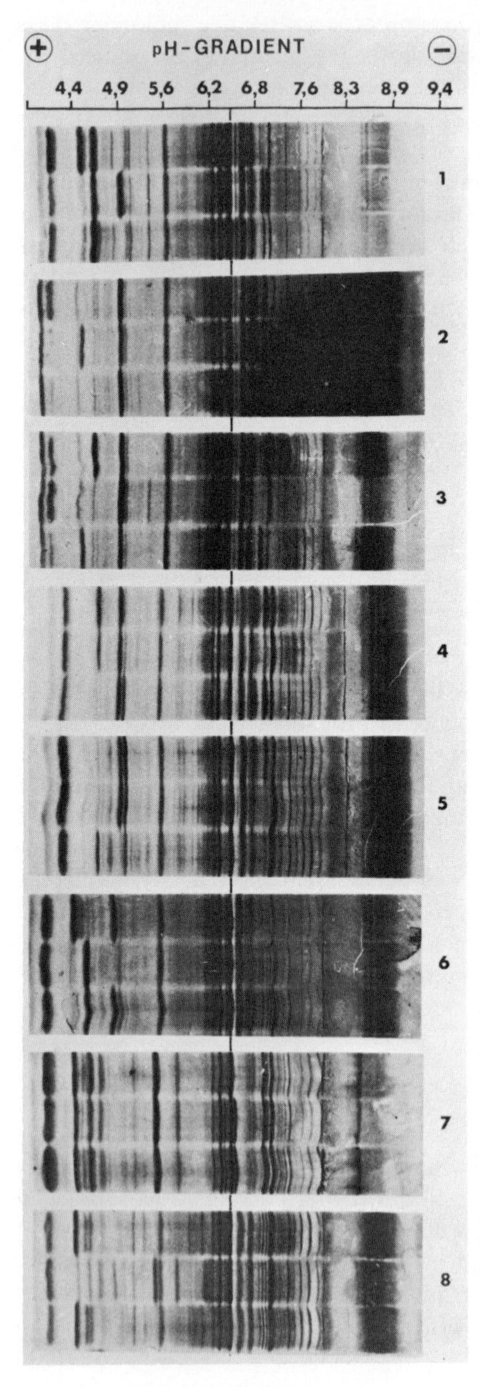

Figure 4.8 White muscle proteins from (1) 'Reeska'; (2) 'Riika'; (3) 'Rääpys'; (4) *Coregonus nasus*; (5) *C. lavaretus*; (6) *C. muksun*; (7) *C. peled*; (8) *C. albula*. Three individual fishes are shown for each species. Nos. 1 to 3 represent dwarfish coregonids from Lake Inari, Finland. In No. 8 the central sample represents an albula × lavaretus hybrid specimen. Samples are arranged along a common central fraction with pI 6.4 to 6.5. Variation in the pH range above 7.3 is slight, while the more acidic parts of the gradient reveal several zones valuable for interspecific comparison. At the extremes of the gradient some interplate variation should be taken into consideration. Reproduced with permission from LKB Instruments.

Figure 4.7 also illustrates the relationship between relative molecular mass and electrophoretic mobility, which allows the technique to be used for estimation of molecular masses. A further technique, which is becoming more widely used with slab electrophoresis, is the use of gradient gels. Here a gradient gel is cast with increasing polyacrylamide concentration, and hence decreasing pore size, in the direction of migration. As the sample moves under the influence of the electric field, the larger molecules will be slowed down and eventually immobilized as they reach a portion of the gel with pore size less than that of the molecular size. Very narrow bands are formed as any trailing molecules will have a chance to 'catch up' as the leading edge of the band is immobilized first. Smaller molecules can then move further through the gel until they too reach a region of appropriate pore size. Gradient-gel electrophoresis provides one of the highest resolution methods for protein separation and has been used to show the presence of some 43 bands in human serum.

The thin-layer configuration also allows two-dimensional electrophoresis to be carried out. The initial electrophoresis is usually carried out in a rod, often as IEF. The rod is then placed on a slab of gel and electrophoresis carried out at right angles to the first direction of migration, the separated components on the rod acting as the sample in the second stage. Complementary separation techniques are used to increase the degree of resolution, with SDS often being used as a second technique to IEF. Two-dimensional electrophoresis allows very detailed protein/peptide maps to be produced which can be used to follow, for example, enzymic hydrolyses of proteins.

Fixing and staining of the gels after electrophoresis is carried out in a similar manner to rod gels, although a number of additional techniques have also been employed. In particular, silver staining (Merill *et al.* 1979) provides a very sensitive means for detecting proteins and nucleic acids. The technique is not suitable for rod gels or agarose gels. A rather different approach to staining the actual gel involves removal of the separated components by 'blotting' onto filter paper or a specially prepared membrane. The separated components may then be more simply detected without interference from the gel matrix by staining, or more commonly, by further immunochemical techniques (Gershoni and Palade 1983).

4.3.6 *Immunoelectrophoresis*

The discriminating power of any separative technique may be greatly enhanced if it is combined with a specific form of detection or visualization and this is the situation encountered in immunoelectrophoresis. A thin layer of agar gel is cast on a glass plate and the sample subjected to electrophoresis. The gel is not stained, but a narrow trough is cut in it parallel to the direction of migration (Figure 4.9). The trough is then filled with the antiserum relevant to the compounds of interest. The gel is left to stand in a moist atmosphere to

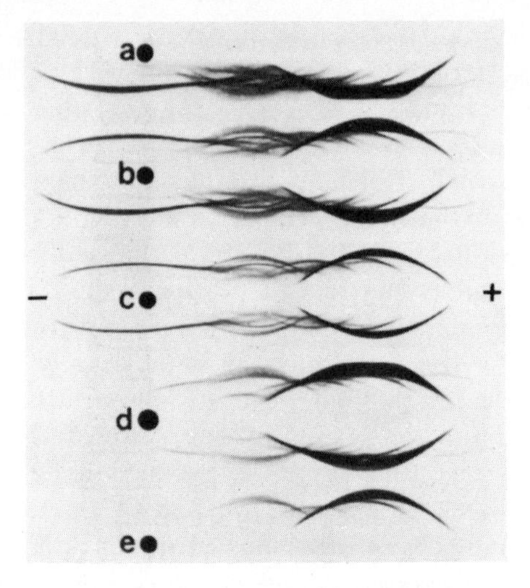

Figure 4.9 Immunoelectrophoresis according to Grabar and Williams. Application wells *a–c*, 2 μl whole human serum in different dilutions (*a*: undiluted, *b*: diluted 1:1, *c*: diluted 1:2) application wells *d–e*, 5 μl and 2 μl human serum with γ-globulins removed by precipitation with ammonium sulphate. Troughs containing antiserum are cut between each well parallel to the direction of migration. Reproduced with permission from LKB Instruments.

allow diffusion of antigen and antibody proteins and immunoprecipitin arcs are formed (Figure 4.9). These are often clearly visible without staining.

An alternative technique involves cutting a strip of the agar gel containing the separated antigens after electrophoresis and then placing this against a second slab of agar containing the relevant antibodies. On electrophoresis at 90° to the original direction, the antigens will migrate into the antibodies and again will form immunoprecipititin arcs, which may be enhanced by suitable staining (Laurell 1966). The characteristic shape of thses peaks has led to the technique being known as 'rocket electrophoresis'. The technique has been widely used for studying human serum proteins and an example is shown in Figure 4.10.

4.4 Isotachophoresis

Isotachophoresis is a slightly different electrophoretic technique in that no support material is used to prevent convection currents within the electrolyte (Everaerts *et al.* 1976). Instead, diffusion is reduced by carrying out the separation in a capillary tube. Furthermore, the effects of dilution caused by electrophoretic migration of buffer ions *through* the sample zones are eliminated by using two quite distinct buffers. The leading buffer contains ions which migrate faster than the sample ions, whereas the terminating buffer contains ions slower than the sample ions, so that the sample ions are never disrupted by buffer ions. The initial and intermediate states of a separation are

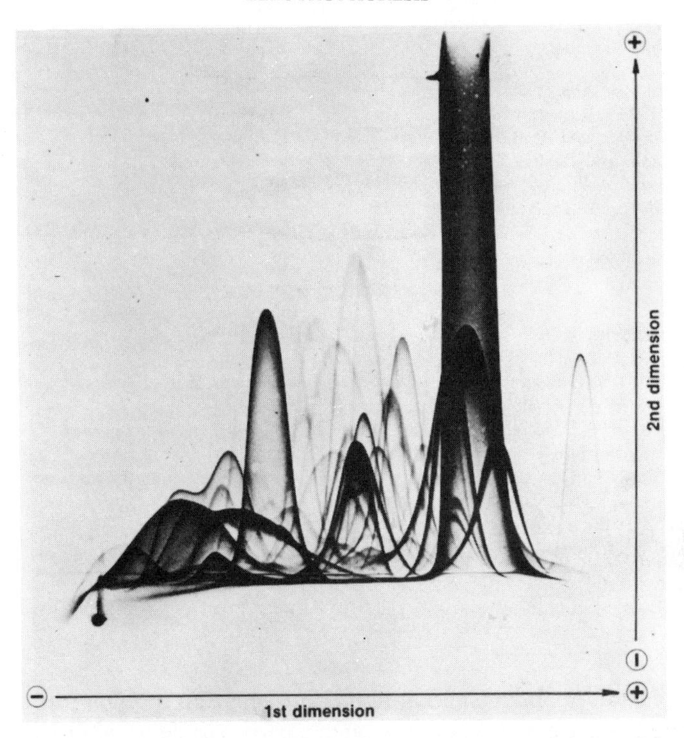

Figure 4.10 Crossed immunoelectrophoresis of 2 μl of human serum. First dimension, separation of sample; second dimension, electrophoresis in an antibody-containing gel. Reproduced with permission from LKB Instruments.

shown in Figure 4.11. The zones are detected by UV absorption or the use of a potential-gradient detector which responds to local changes in the electric field as sample zones pass through the system.

The technique has found wide application for small organic and inorganic ions, for example inorganic ions in waste water (NO_3^-, SO_4^{2-}) or amino acids and organic acids in wines, fruit juices and physiological samples.

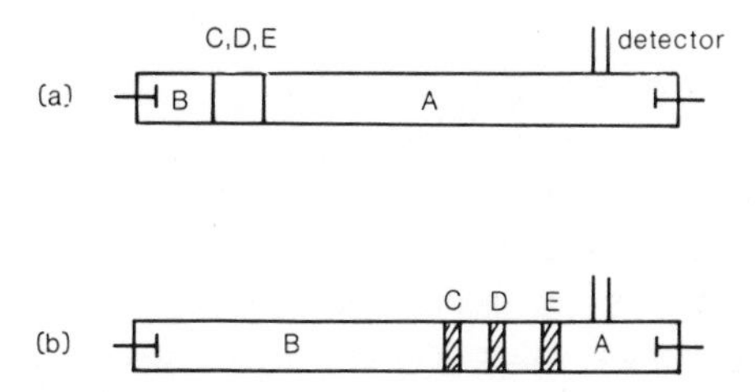

Figure 4.11 Isotachophoresis. (*a*) A sample containing three ions (*C*, *D* and *E*) is introduced between the leading (*A*) and terminating (*B*) buffers. (*b*) Current is applied between electrodes, and ions migrate, eventually separating.

References

Davis, B.J. (1964) *Ann. N.Y. Acad. Sci.* **121**, 404.

Everaerts, F.M., Beckers, J.L. and Verheggen, T.P.E.M. (1976) *Isotachophoresis—Theory, Instrumentation and Applications.* Elsevier Science Publishers, New York.

Gershoni, J.M. and Palade, G.E. (1983) *Analyt. Biochem.* **124**, 396.

Grabar, P. and Williams, C.A. (1953) *Biochim. Biophys. Acta.* **10**, 193.

Laurell, C.B. (1966) *Analyt. Biochem.* **15**, 45.

Merill, C.R., Switzer, R.C. and Van Keuren, M.L. (1979) *Proc. Nat. Acad. Sci. USA* **76**, 4335.

Ornstein, L. (1964) *Ann. N.Y. Acad. Sci.* **121**, 321.

Further reading

Andrews, A.T. (1981) *Electrophoresis: Theory, Techniques and Biochemical and Clinical Applications*, Oxford University Press, Oxford.

Deyl, Z. (1980–1982) *Electrophoresis: A Survey of Techniques and Applications. Parts A and B.* Elsevier, Amsterdam.

Goal, O., Medgyesi, G.A. and Vereczkey, L. (1980) *Electrophoresis in the Separation of Biological Macromolecules*, Wiley-Interscience, New York.

Smith, I. and Seakins, J.W.T. (1976) *Chromatographic and Electrophoretic Techniques*, Vols. I and II, 4th edn., Heinemann Medical, London.

5 Introduction to spectroscopy

5.1 Spectroscopy

Spectroscopy involves the measurement of the absorption and emission of radiation by atoms and molecules and the interpretation of these studies in order to gain information about the structure or concentration of the species interacting with the radiation. Spectroscopic techniques have developed dramatically in the past forty years and have facilitated great advances in both the physical and biological sciences.

5.2 The electromagnetic spectrum

Radiation has characteristics consistent with both an electromagnetic waveform and a beam of particles or photons. It is said to be monochromatic if all the photons have the same energy, which corresponds to light of a single wavelength. Radiation induces transitions between discrete energy levels in an atom or molecule if the wavelength λ of the radiation is related to the energy difference between two levels ΔE by the equation

$$\Delta E = \frac{hc}{\lambda}$$

where h is the Planck constant and c is the velocity of light in a vacuum; ΔE is described as a quantum of energy with units of $\mathrm{J\,mol^{-1}}$. It is also related to the frequency ν (in units of Hz), since

$$\nu = \frac{c}{\lambda}$$

Wave numbers $\bar{\nu}$ are used to express frequency in IR spectroscopy. They are the reciprocal of the wavelength, with units of $\mathrm{cm^{-1}}$.

Molecules have discrete nuclear spin, electron spin, rotational, vibrational and electronic energy levels. Radiation will be absorbed and cause transitions between appropriate energy levels depending on its frequency (Figure 5.1). Nuclear magnetic resonance (NMR), electron spin resonance (ESR), IR, visible and UV spectroscopy are the major procedures for studying molecular absorption of radiation in the appropriate regions of the electromagnetic spectrum. Fluorescence and phosphorescence involve studies of the emission of radiation after initial absorption of radiation. Atomic absorption and atomic emission spectroscopy are procedures for studying electronic transitions in atoms.

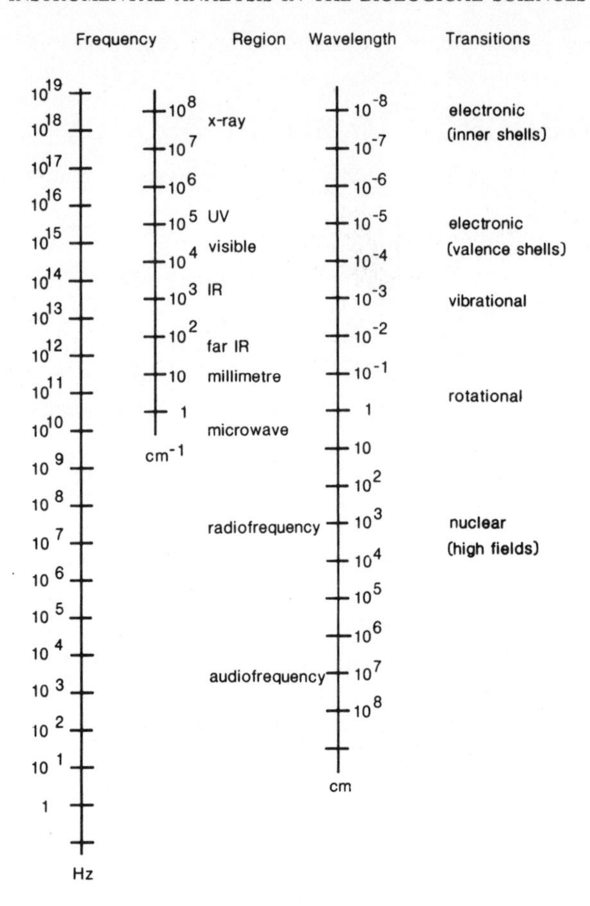

Figure 5.1 Regions of the electromagnetic spectrum and corresponding molecular transitions.

5.3 Molecular energy states

The energy of a particular mode of molecular motion is assigned a value of zero when the molecule is in the ground state. This is the state which is occupied to a greater extent as the material is cooled slowly towards absolute zero. The excited states of a molecule are the higher energy levels to which the molecule can be excited by the absorption of radiation. If two or more energy states are equal in energy, they are said to be degenerate. Degenerate energy states can often be split by the application of a weak external influence, such as a magnetic field.

At thermal equilibrium at a temperature TK, atoms and molecules will be distributed among the available energy levels according to the Boltzmann distribution. The relative probability P of a molecule being in the ith energy state, with an energy E_i above the ground state is given by

$$P = \frac{\exp(-E_i/kT)}{\sum_i \exp(-E_i/kT)}$$

where k is the Boltzmann constant and $\sum_i$ represents the sum of the exponential term over all the energy levels.

The energy of an electronic state varies with interatomic distance, as shown in Figure 5.2 for a diatomic molecule. The difference in energy between electronic states is greater than that between vibrational states, while the energy difference between rotational states is even smaller. The Franck–Condon principle states that the absorption of radiation is rapid compared

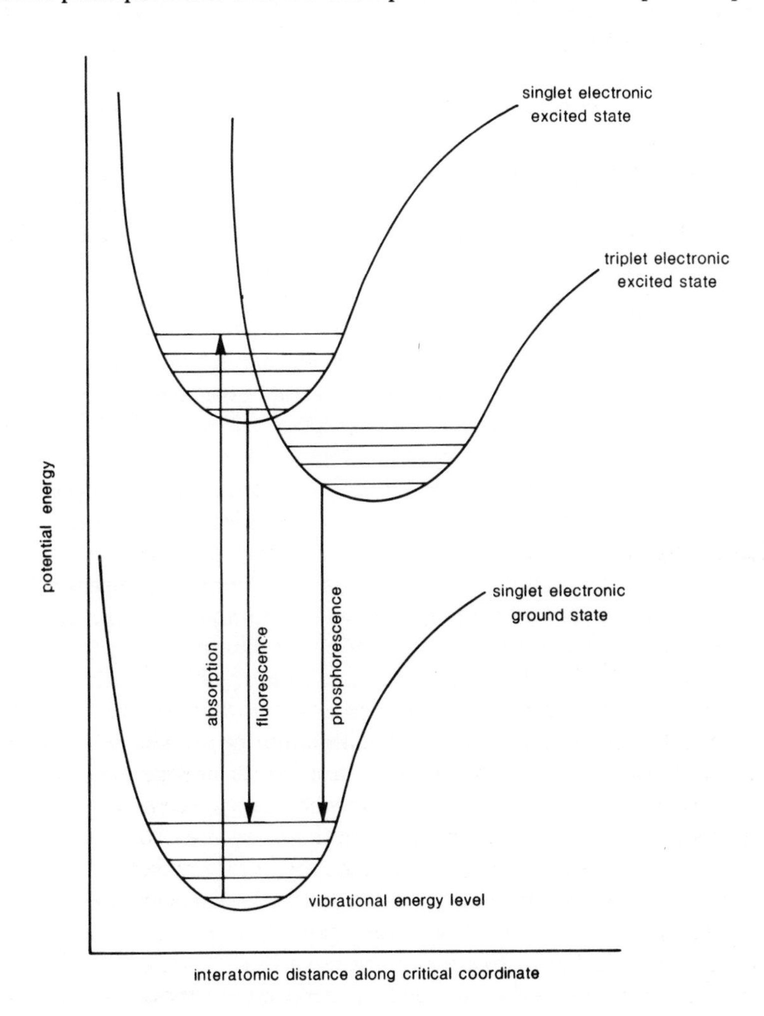

Figure 5.2 Potential energy diagram for a diatomic molecule.

with nuclear motion, and therefore atomic and molecular transitions can be represented by vertical lines in an electronic energy diagram of this type.

Since the energy levels of a molecule are dependent on its structure, spectroscopic techniques are suitable for structural determination and identification of molecules. The simple relationship between intensity of absorption and component concentration (see Section 5.5) allows spectroscopic studies to be extended to the quantitative determination of molecules.

5.4 Molecular transitions

Absorption of radiation occurs as a molecule is excited from a lower to a higher energy state, while emission occurs as a molecule emits a quantum of energy in moving from a higher to a lower energy state. The probability of the absorption process is proportional to the intensity of the radiation of the required frequency and to the number of molecules in the lower energy state and the probability of spontaneous emission is proportional to the number of molecules in the upper state. Emission induced by the radiation also occurs with a probability proportional to the number of molecules in the upper state and the intensity of the radiation of the appropriate frequency.

Emission and absorption processes both occur during a spectroscopic study. A net absorption of radiation of a particular frequency will occur if the number of molecules in the lower energy state exceeds that in the upper state. If the absorption of radiation leads to equal numbers of molecules in the upper and lower energy states, there is no net absorption of radiation. If the signal disappears it is described as saturated. This applies to techniques such as NMR spectroscopy where all radiation emitted reduces the absorption signal. However, in techniques such as UV or visible spectrophotometry, the radiation is absorbed from a focused beam, while emission of radiation occurs in all directions. Hence, the detector monitoring the emission of radiation in a single direction will perceive an absorption of radiation along the beam axis, even though the net absorption of radiation by the sample is zero.

Molecules can return to the ground state from the excited state either by emission of radiation (equally in all directions) or without the emission of radiation by exchanging energy with neighbouring molecules. The latter process is important in preventing saturation in techniques where the difference in population of the lower and higher energy states is small, such as NMR spectroscopy. Non-radiative transitions are also important in fluorescence and phosphorescence spectroscopy. Fluorescence involves the absorption of radiation followed by the loss of vibrational energy by a non-radiative process prior to emission of radiation. Phosphorescence involves an electronic transition prior to emission of radiation.

The radiation absorbed or emitted during a molecular transition is only monochromatic if the molecular energy levels are sharply defined. However.

according to the Heisenberg uncertainty principle, the energy of a system at a particular time cannot be precisely defined. Therefore the energy of a transition has a significant uncertainty, which is reflected in the range of radiation frequencies absorbed or emitted. This leads to the broadening of spectral peaks which is increased by instrumental limitations and the fact that various processes such as electronic, vibrational and rotational excitation may occur together and not be resolvable. Each peak is characterized by the wavelength (or frequency) of maximum absorption, which is denoted as λ_{max} in UV–visible spectrophotometry.

5.5 Quantitative analysis

Quantitative analysis in photometric absorption spectroscopy is based on the Beer–Lambert law. This law is a combination of Lambert's law which relates the radiation absorbed to the cell pathlength, and Beer's law which relates the radiation absorbed to the solute concentration.

Lambert's law states that the change in radiation intensity, dI_x, as it passes through a thin section of a transparent medium is proportional to the incident radiation intensity I_x and the distance traversed by the beam dx. The arrangement considered is illustrated in Figure 5.3, and the law may be represented mathematically as

$$-dI_x = kI_x dx \tag{5.1}$$

where k is a constant of proportionality.

Beer's law states that each molecule of solute absorbs the same fraction of radiation incident upon it, regardless of concentration, in a non-absorbing medium. Thus the constant k in (5.1) is proportional to the number of molecules, or the molar concentration of solute, C.

$$k = aC \tag{5.2}$$

where a is a constant of proportionality.

Equations (5.1) and (5.2) can be combined into the Beer–Lambert equation:

$$-dI_x = aCI_x dx \tag{5.3}$$

This can be integrated between $x = 0$ and $x = l$, where l is the cell pathlength.

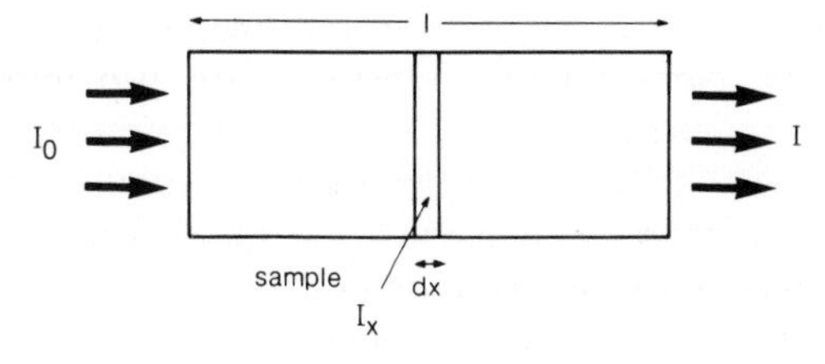

Figure 5.3 Diagram for deriving Lambert's law.

When $x = 0, I_x = I_0$ which is the intensity of the incident radiation, and when $x = l, I_x = I$ which is the intensity of the transmitted radiation at the required wavelength.

$$\int_{I_0}^{I} \frac{-\mathrm{d}I_x}{I_x} = \int_0^l aC\,\mathrm{d}x$$

Thus

$$\ln I_0/I = aCl$$

or

$$I/I_0 = \exp(-aCl) \tag{5.4}$$

The left-hand side of (5.4) is the fraction of radiation transmitted by the solution. When C is in mol dm^{-3}, l in cm, and logarithms to the base 10 are used, the proportionality constant is denoted by ε and is known as the molar absorptivity or molar absorption coefficient. Older texts describe ε as the molar extinction coefficient or the molar absorbancy index. The absorbance of a sample is defined as $\log_{10} I_0/I$, and I/I_0 is called the transmittance. When the relative molecular mass of a substance is unknown, the intensity of absorption in the UV–visible region is expressed as the $E_{1\,\mathrm{cm}}^{1\%}$ value, which is the absorbance of a 1% solution in a 1 cm cell.

The Beer-Lambert Law is often followed by dilute solutions studied by IR, UV and visible spectroscopy. It also forms the basis for quantitative studies in fluorescence and phosphorescence spectrophotometry. Deviations from Beer's law may occur because of the association of solute molecules or because of changes in the RI of the solution at higher concentrations. Shifts in the position of chemical or physical equilibria with increases in solute concentration may also lead to a deviation from linearity. Beer's law will also fail if the absorbance of the solute varies over the range of wavelengths incident on

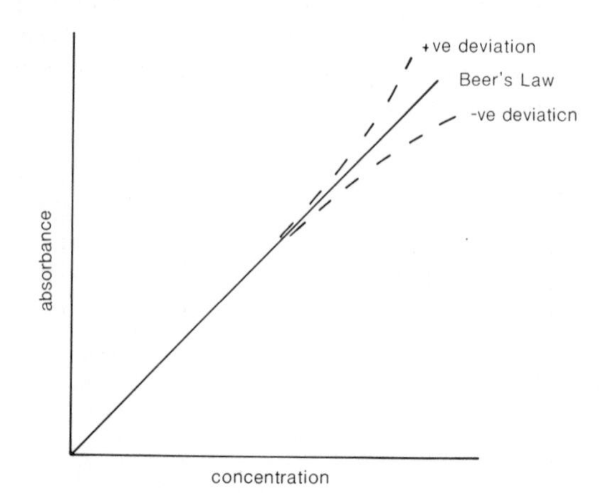

Figure 5.4 Representation of Beer's law showing possible deviations.

the sample. This is generally not a problem with spectrometers where the monochromators only transmit a narrow range of wavelengths, but it can become a problem with colorimeters where filters transmit a relatively broad band of radiation.

Since positive or negative deviations from Beer's law may occur, as shown in Figure 5.4, it is essential that standards covering the concentration range of interest are used to draw up a calibration curve for quantitative studies.

5.6 Determination of a spectrum

The absorption spectrum of a molecule is the plot of the fractional amount of radiation absorbed at each frequency (or wavelength) as a function of frequency (or wavelength). Instruments used to determine the spectrum are called spectrophotometers. This term is precise, since it indicates that the instrument determines the intensity of absorption, but it is often abbreviated to spectrometer, except for instruments for which at least part of the frequency region scanned is in the visible. The technique used to measure a spectrum is usually termed spectroscopy, although spectrometry is used for the technique of mass spectrometry. The quantitative analysis of chromophores in the visible and ultraviolet region is referred to as spectrophotometry.

Spectrometers have several features in common (Figure 5.5). For absorption spectroscopy, a radiation source is required and the radiation must be focused and converted into a narrow beam with collimating slits or a collimating lens. Sources other than lasers emit a broad band of radiation and a monochromator is required to filter out all radiation except that at a single wavelength. In practice, monochromators transmit light over a narrow range of wavelengths and the narrowness of this range is the major factor affecting the resolving power or resolution of the instrument. After passing through the monochromator, the radiation is transmitted or reflected by a sample. A collimating slit (or collimator) may be required to remove scattered radiation and the beam then passes to the detector, which generates an electrical signal

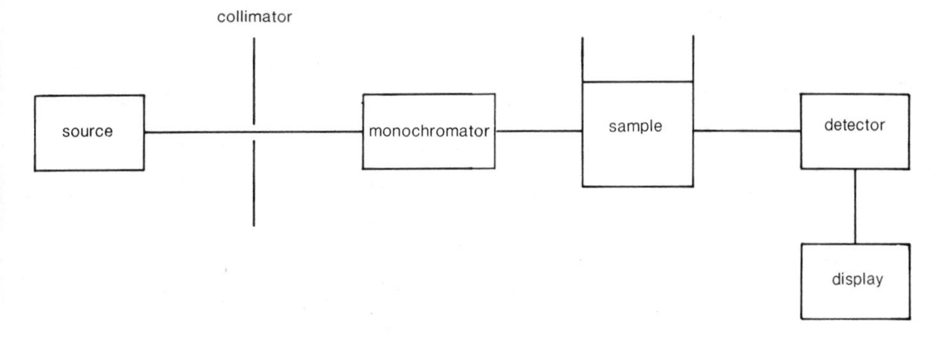

Figure 5.5 Schematic diagram of a single-beam absorption spectrometer.

proportional to the intensity of the radiation incident upon it. The electrical signal is recorded by a chart recorder or stored in a computer. If there is no absorption or emission of radiation by the sample, the signal represents the baseline. Instrumental limitations may give rise to noise which is observed as rapid fluctuations in the baseline. A monochromator is located between the sample and the detector in instruments for emission studies such as fluorescence, phosphorescence and atomic emission spectroscopy.

The single-beam spectrometer shown in Figure 5.5 is generally limited to measurements at a given wavelength, since changes in source intensity and detector sensitivity with wavelength preclude the use of this design in instruments that scan a range of wavelengths. In a single-beam instrument, quantitative measurements can be performed by selecting a wavelength, which is usually the absorption maximum of the solute, and placing the reference solution, which is generally the solvent, in the radiation beam. The instrument is adjusted to read 100% transmittance (zero absorbance) and the reference cuvette is replaced by the sample. The absorbance is then read and can be related to the concentration of solute with a calibration curve.

Double-beam instruments are employed as scanning spectrometers. Radiation from the source is passed through a monochromator and the beam is then split into two components of equal intensity (Figure 5.6). One beam passes through the sample cell and the second beam passes through the reference cell. The difference in the electrical signal produced by the two beams at the detector is converted into the absorbance value, which is displayed. The double-beam design eliminates any effects due to absorption of radiation by the atmosphere (which is particularly important in IR spectroscopy) or by the solvent, and it minimizes effects due to source instability or amplifier drift. Instruments are generally designed with a mechanism for maintaining the incident radiation at a constant level as the spectrum is recorded. This may involve electronic circuitry to regulate the detector sensitivity via a feedback loop (known as automatic gain control); the slit width may be controlled automatically; or the position of an optical wedge in the light path may be controlled to increase or attenuate the level of radiation reaching the detector.

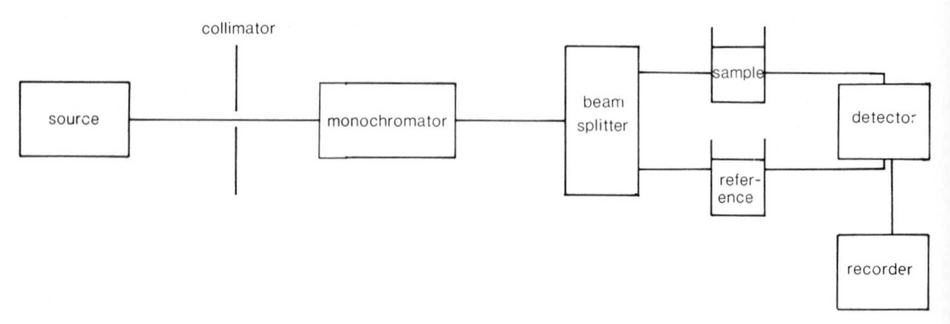

Figure 5.6 Schematic diagram of a double-beam spectrometer.

The splitting of the radiation in a double-beam spectrometer may be achieved simply by positioning a mirror in the light path so that half the radiation is deflected through a reference cell and half passes through to the sample cell. However, more commonly, the optical beam is directed alternately through the sample and the reference cells by a system of rotating sector mirrors known as choppers. The open sector allows the beam to pass through for focusing through the sample on to one channel of the detector, while the closed sector reflects the radiation through the reference onto a second channel of the detector.

Further reading

Brown, S.B. (1980) *An Introduction to Spectroscopy for Biochemists*, Academic Press, London.
Crooks, J.E. (1978) *The Spectrum in Chemistry*, Academic Press, London.

6 UV–visible spectrophotometry

6.1 Introduction

The visible region of the spectrum ranges from 400–700 nm, which corresponds to frequencies of 7×10^{14}–4×10^{14} Hz. Radiation at the low-energy high-wavelength end of the spectrum is perceived as red, while that at the high-energy part of the spectrum is blue. The spectral region between 180–400 nm corresponds to the ultraviolet (UV) region.

The human eye detects light transmitted through or reflected from samples. Therefore, the colour of a sample represents the complementary colour to that which is absorbed. Thus, a sample which appears blue will absorb light at the red end of the spectrum when studied in a spectrophotometer. Absorption of light in the UV or visible region causes the excitation of a valence electron into the lowest energy unoccupied molecular orbital. Since the radiation stimulating electronic transitions is high in frequency, and hence energy, compared to that required to induce vibrational and rotational transitions, the absorption of light in this region of the spectrum may cause vibrational and rotational

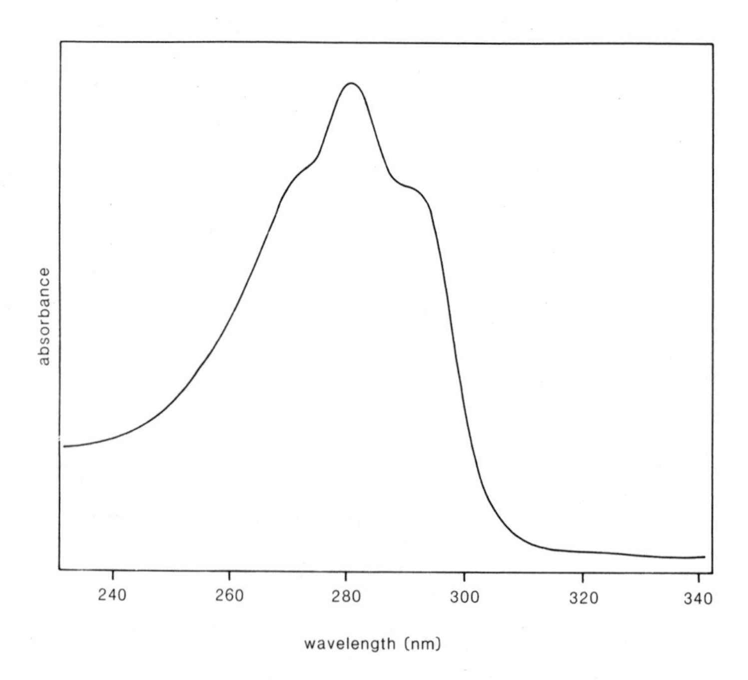

Figure 6.1 Ultraviolet spectrum of leukotriene D_4 in methanol.

excitation as well as electronic excitation. UV and visible spectra of solutions often show partially resolved or more commonly broad absorption bands due to incomplete resolution of the individual absorption peaks corresponding to different combinations of vibrational and rotational transitions accompanying the electronic excitation.

A typical UV spectrum is that of leukotriene D_4, shown in Figure 6.1. As a consequence of the broad bands observed in this branch of spectroscopy and the limited amount of structural information conveyed by the spectrum, UV and visible studies are frequently concerned with quantitative measurements rather than structural identification of unknown compounds. However, the wavelengths of maximum absorption (λ_{max}) may provide useful information about the structure of an unknown compound which complements that obtained from other sources.

6.2 Electronic energy levels

In order to understand electronic transitions, some knowledge of molecular orbital theory is required and a basic treatment is included here. The electrons

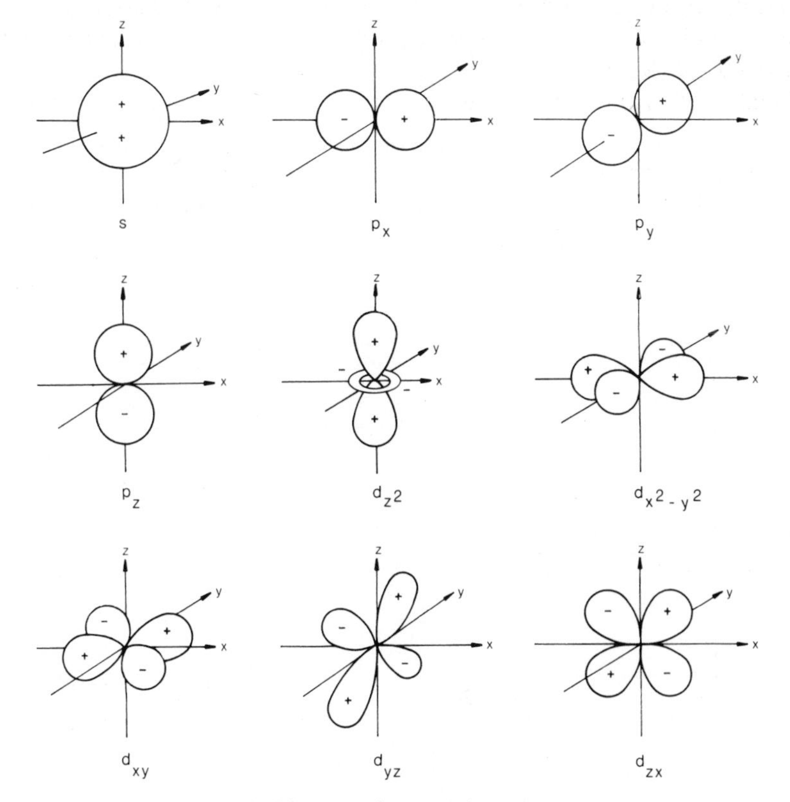

Figure 6.2 Diagrams of *s*, *p* and *d* atomic orbitals.

of an atom in its ground state occupy the lowest available energy levels or orbitals. According to the Pauli exclusion principle, a maximum of two electrons with opposed spins are allowed in each orbital. If two equal energy (degenerate) orbitals are available for two electrons, the lowest energy state corresponds to that in which one electron is located in each orbital with parallel spin according to Hund's rule. The available orbitals vary in symmetry, being designated by *s*, *p*, *d* and *f* (Figure 6.2), and they are distributed in shells according to their energy. Thus, an iron atom has the electronic configuration $1s^2\ 2s^2\ 2p^6\ 3s^2\ 3p^6\ 3d^6\ 4s^2$. In shells where *p* and *d* orbitals are available, there are three *p* and five *d* orbitals. The symmetry of atomic orbitals is reflected in that of the molecular orbitals formed by two atoms. Either σ, π or non-bonding (n) molecular orbitals may be occupied. A σ molecular orbital is symmetrical about the internuclear axis between the two atoms forming the bond, while the two lobes of a π orbital have opposite sign. To a first approximation, the molecular orbitals can be considered to arise from a linear combination of two atomic orbitals which are similar in energy. However, mixing or hybridization of the *s* and *p* atomic orbitals may occur to give sp, sp^2 or sp^3 atomic orbitals which can overlap to a greater extent with the orbitals of the second atom, thereby giving a lower energy molecular orbital. Overlap of two *s* atomic orbitals gives rise to a σ molecular orbital and overlap of two *p* atomic orbitals may give rise to a σ or a π molecular orbital, depending on their orientation (Figure 6.3). If an atomic orbital is not similar in energy to that of an orbital on the second atom it cannot contribute to the bonding and it remains a non-bonding molecular orbital.

A combination of two atomic orbitals gives rise to a low energy or bonding molecular orbital, denoted by σ or π, and a higher-energy, antibonding orbital

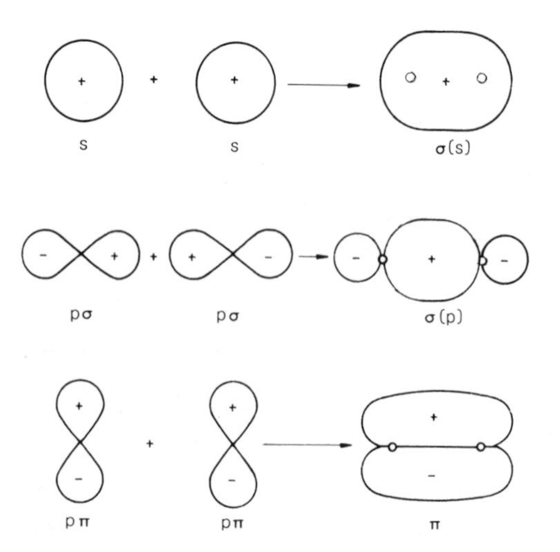

Figure 6.3 Diagrams illustrating the formation of some simple two-centre molecular orbitals from atomic orbitals.

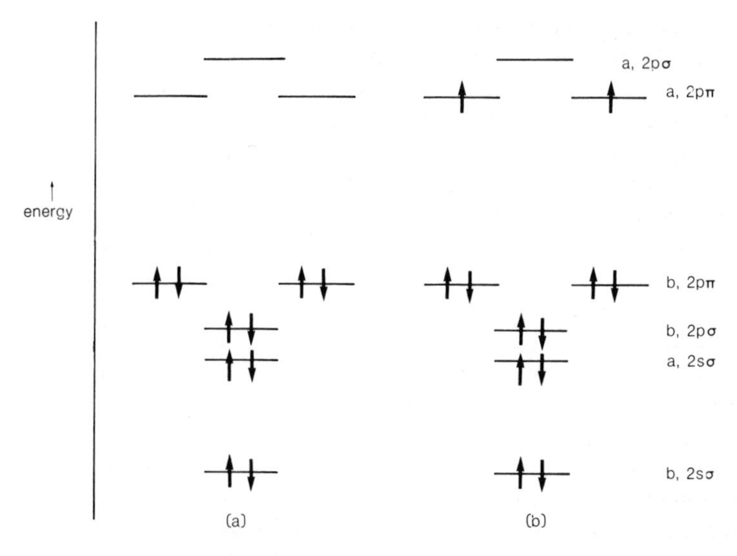

Figure 6.4 Molecular structures: (*a*) nitrogen in its singlet ground state; and (*b*) oxygen in its triplet ground state.

denoted by σ^* or π^*. The location of electrons in the molecular orbitals of nitrogen and oxygen molecules in their ground states is shown in Figure 6.4. In general, the energy of molecular orbitals increases in the order $\sigma < \pi < n < \pi^* < \sigma^*$. Electronic structures, such as the ground state of nitrogen, which have all electron spins paired, are called singlet forms, while configurations in which two unpaired electrons occur with parallel spins, as in the ground state of oxygen, are known as triplet states. Most molecules have a singlet configuration in the ground state.

6.3 Electronic transitions

UV and visible light induces transitions between electronic energy levels. However, not all transitions are equally probable. Certain selection rules provide guidelines as to which transitions are allowed or forbidden. Transitions between singlet and triplet states involve a change in the spin angular momentum of the molecule and are strictly forbidden. However, other transitions may occur, despite being forbidden, due to an interaction between electronic and vibrational energy levels, although the intensity of the absorption is often low. The $n \rightarrow \pi^*$ transition of a carbonyl group belongs to this category. The molar absorption coefficient is $< 100\ M^{-1}\ cm^{-1}$ compared with up to $100\,000\ M^{-1}\ cm^{-1}$ for some allowed transitions. The electronic transitions observed in the UV and visible region of the spectrum in order of decreasing energy are $\sigma \rightarrow \sigma^* > n \rightarrow \sigma^* > \pi \rightarrow \pi^* > n \rightarrow \pi^*$. The energy required

Table 6.1 Spectral parameters of common chromophores

Chromophore	*Example*	λ_{max} (nm)	ε	*Solvent*
>C=C<	Ethylene	171	15 530	Vapour
>C=O	Acetone	166	16 000	Vapour
		189	900	Hexane
		279	15	Hexane
$—CO_2H$	Acetic acid	208	32	Ethanol
—C≡N	Acetonitrile	167	Weak	Vapour
$—CONH_2$	Acetamide	178	9500	Hexane
		220	63	Water

for a $\sigma \rightarrow \sigma^*$ transition is very high and is usually beyond the UV region for most molecules.

Compounds that contain non-bonding electrons on oxygen, nitrogen, sulphur or halogen atoms may show absorption peaks due to $n \rightarrow \sigma^*$ transitions. Transitions to π^* orbitals are associated with unsaturated groups in a molecule. Saturated aldehydes and ketones have a low-intensity absorption around 285 nm due to an $n \rightarrow \pi^*$ transition as well as a high-intensity absorption due to a $\pi \rightarrow \pi^*$ transition at about 180 nm, which is outside the normal UV range.

A functional group which absorbs UV or visible radiation is called a chromophore. The wavelength corresponding to maximum absorption λ_{max} and the molar absorption coefficient ε of some common chromophores are given in Table 6.1. Saturated hydrocarbons, or saturated molecules containing hydroxyl or ether groups have no absorption above 180 nm. Compounds of this type, including hexane, ethanol, diethyl ether and water, are commonly used as solvents for the determination of UV and visible spectra. Substituents which are not themselves chromophores may modify the absorption of chromophores. Substituents of this type, including $—CH_3$, —Cl, $—NH_2$ and —OH are termed auxochromes.

6.3.1 *Solvent effects*

Electronic transitions involve a redistribution of electrons within a molecule and consequently the polarity of the excited state often differs from that of the ground state. Since polar solvents will stabilize a polar electronic state to a greater extent than non-polar solvents, there is often a change in λ_{max} and ε with a change of solvent. Solvent effects and the effects of auxochromes are described by the following terms:

(a) hyperchromic: an increase in intensity of absorption
(b) hypochromic: a decrease in intensity of absorption

(c) bathochromic: a shift to longer wavelength
(d) hypsochromic: a shift to shorter wavelength

A polar solvent may cause a bathochromic shift if the excited state is more polar than the ground state. Bands corresponding to $\pi \rightarrow \pi^*$ transitions sometimes show a small solvent effect of this type. However, solvent effects cause greater shifts for bands corresponding to $n \rightarrow \pi^*$ and $n \rightarrow \sigma^*$ transitions. A hypsochromic shift of about 11 nm occurs for an α, β-unsaturated ketone on changing from an aliphatic hydrocarbon to ethanol as solvent, reflecting the lower polarity of the excited state relative to the ground state.

6.3.2 *Effect of conjugation*

When two or more double bonds are separated from each other by a single bond, they are said to be conjugated. In molecules of this type, the π electrons are not restricted to the formal double bonds but are delocalized across all the atoms of the conjugated system. This causes a reduction in the energy of the $\pi \rightarrow \pi^*$ transition and a bathochromic shift is observed. The molar absorption coefficient is also increased. The effect of conjugation can be readily understood by comparing the molecular orbitals of two isolated ethylene molecules with those of the conjugated diene, butadiene, as shown in Figure 6.5. The π orbitals of the two ethylene molecules are split into a higher and lower energy π orbital in butadiene. Similarly, the π^* orbitals of the ethylene molecules are split into a higher and lower energy π^* orbital. Hence, the energy of the $\pi_1 \rightarrow \pi_1^*$ transition of butadiene is significantly lower than the $\pi \rightarrow \pi^*$ transition of ethylene, and λ_{max} for butadiene occurs at 217 nm compared with 171 nm for ethylene.

The greater the number of double bonds in the conjugated system, the lower

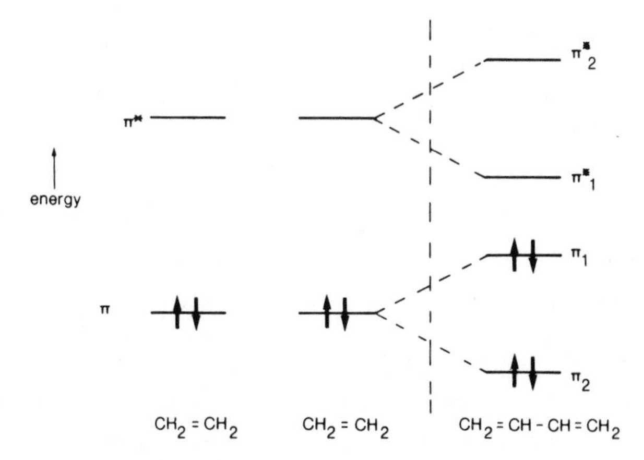

Figure 6.5 Effect of conjugation of the π orbitals of two ethylene molecules on the molecular energy levels in butadiene.

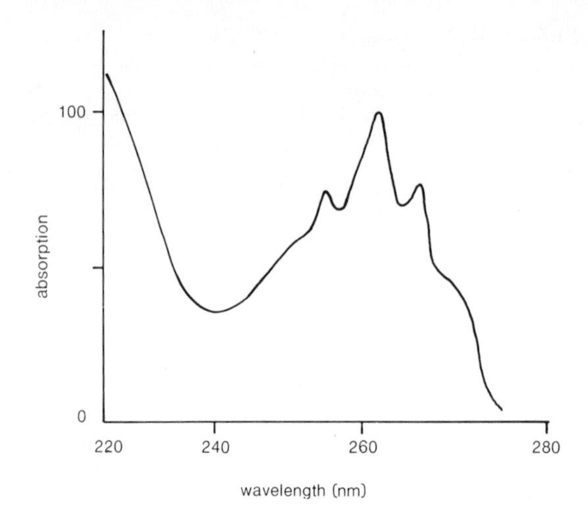

Figure 6.6 Ultraviolet spectrum of phenylalanine at pH 6.

the energy of the $\pi \rightarrow \pi^*$ transition and the higher the molar absorption coefficient of the molecule. If the number of conjugated double bonds is sufficient, the absorption will be shifted into the visible part of the spectrum and the compound will appear coloured. For hydrocarbons, seven conjugated double bonds are required for the compound to be coloured. Thus, for β-carotene with eleven conjugated double bonds, λ_{max} is at 450 nm and the compound is orange in colour. Delocalization of electrons can only occur in a conjugated molecule if the atoms are coplanar. In some molecules, steric effects force some of the atoms out of the plane and prevent delocalization. Aromatic compounds show several weak absorption bands associated with vibrational effects on the $\pi \rightarrow \pi^*$ transition. This is illustrated by the spectrum of the aromatic amino acid, phenylalanine (Figure 6.6).

6.4 Qualitative analysis

The value of λ_{max} is dependent on the structure close to the unsaturated groups of a molecule. The major factors affecting λ_{max} are the nature of the double bonds ($>C=C<$ $>C=O$, etc.), the number of conjugated double bonds and the presence of substituents or rings at the unsaturated groups. Simple rules can be used to predict λ_{max} (Table 6.2) and the predicted values agree well with the observed λ_{max} values in many cases, as shown in Table 6.3.

These rules are only useful for relatively simple conjugated structures, but often the comparison of the absorption spectrum of a molecule with that quoted in the literature provides valuable structural confirmation and may

Table 6.2 Effects of structure on λ_{max} values

Class of compound	*Structural component*	λ_{max}	*Shift in* λ_{max}
Acylic diene	—	217	
Homoannular diene	—	253	
Diene	Alkyl substituent		+5
Diene	Cl, Br		+17
Diene	Exocyclic $>C=C<$		+5
Diene	Extra conjugated $>C=C<$		+30
α, β-unsaturated compound	—	215	
α, β-unsaturated compound	α-alkyl substituent		+10
α, β-unsaturated compound	β-alkyl substituent		+12
α, β-unsaturated compound	γ- or δ-substituent		+18
α, β-unsaturated compound	Extra conjugated $>C=C<$		+30
α, β-unsaturated compound	Enolic α-OH		−35
α, β-unsaturated compound	Enolic β-OH		−35
α, β-unsaturated compound	α-Cl		−15
α, β-unsaturated compound	α-Br		−23

Table 6.3. Predicted and observed λ_{max} values

Compound	*Predicted* λ_{max} (nm)	*Observed* λ_{max} (nm)
Cholesta-2,4-diene	273	276
Cholesta-3,5-diene	237	235
Cholesta-5,7-dien-3β-ol	263	262
1-Cholesten-3-one	227	231
4-Cholesten-3-one	239	241
Cholesta-3,5-dien-7-one	275	278
3β-Hydroxycholest-5-en-7-one	239	237

also reveal impurities. Thus, the spectra of chlorophyll a and chlorophyll b (Figure 6.7) are sufficiently different to confirm the structure and determine whether there is any contamination of one pigment by the other.

The major limitations in applying UV and visible spectrophotometry in qualitative analysis arise from the fact that large numbers of molecules do not show adequate absorption and, in addition, the spectrum only reflects the local structure in the vicinity of the double bonds. Many molecules with widely differing structures show the same λ_{max} value.

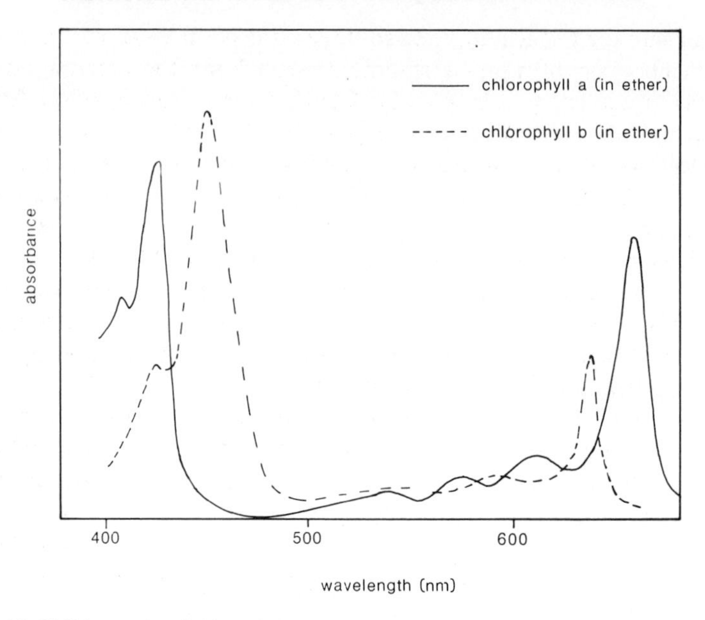

Figure 6.7 Visible spectra of chlorophyll a and chlorophyll b in ether. Redrawn with permission from Goodwin (1983).

6.5 Quantitative analysis

The main application of UV and visible absorption studies is in quantitative analysis. The absorption of light in this region of the spectrum is governed by the Beer–Lambert law, $\log(I_0/I) = \varepsilon cl$, as described in Chapter 5. It is preferable to determine a calibration curve, rather than relying on values of the molar absorption coefficient reported in the literature, so that any deviations from Beer's law or other effects, such as non-linearity of the detector, can be allowed for in subsequent determinations. Molar absorption coefficients for electronic transitions are high, with values of 20 000–30 000 $M^{-1}\,cm^{-1}$ for molecules containing two conjugated double bonds and values up to 100 000 $M^{-1}\,cm^{-1}$ for highly conjugated molecules. This contrasts with the IR region where maximum values of about 1 000 are obtained. Thus, UV and visible spectrophotometry is a very sensitive analytical procedure.

Since UV–visible spectrophotometry is relatively unspecific, separation of components in a mixture is often required before quantitative determination. The use of a liquid chromatographic technique, particularly HPLC, prior to quantification is very common. Indeed, a UV–visible detector is the most common method of detection and quantification in HPLC (see Section 2.4.4). An alternative method of extending the range of application of UV–visible spectrophotometry for quantitative analysis involves the use of a chemical or biochemical reaction, either to shift the absorption of a non-absorbing molecule into the UV region, or to shift the absorption away from interfering components. For example, a standard method for the determination of polyunsaturated fatty acids in edible fats involves treatment of the fat with potassium hydroxide solution at 180 °C (IUPAC 1979). This causes the isomerization of the naturally occurring methylene-interrupted polyunsaturated fatty acids, which do not absorb in the UV, to conjugated dienoic and trienoic fatty acids, which absorb at 233 nm and 268 nm respectively.

Changes in the absorbance of a sample in the visible or UV region are a useful way of studying the kinetics of molecular reactions and assaying the activity of specific enzymes. Often the assay is carried out in the cuvette of the spectrophotometer, with the change in absorbance at a given wavelength with time being recorded directly on a chart recorder. For example, the lipoxygenase activity in a plant can be determined by monitoring the absorbance of an emulsion containing 9,12-octadecadienoic acid, enzyme solution and buffer at 234 nm. The enzyme catalyses the formation of conjugated diene hydroperoxides and the enzyme activity can be defined as the maximum rate of change in absorbance per millilitre of enzyme solution (St Angelo and Ory 1972).

Often the enzyme of interest does not produce a product which absorbs in the UV or visible region or there is interference by other components at the product λ_{max}. In these cases, the enzyme activity can be determined if the primary reaction is coupled with a fast secondary reaction. The assay of the enzyme hexokinase (HK), which is involved in glycolysis, is a typical example of a coupled assay. The D-glucose-6-phosphate produced by the primary reaction reduces $NADP^+$ to NADPH in the presence of excess glucose-6-phosphate dehydrogenase (G6P-DH), and formation of NADPH can be monitored by spectrophotometry at 340 nm. The reactions involved are as follows:

$$\text{D-glucose} + \text{ATP} \underset{}{\overset{\text{HK}}{\rightleftharpoons}} \text{ADP} + \text{D-glucose-6-P}$$
$$\text{D-glucose-6-P} + \text{NADP}^+ \underset{}{\overset{\text{G6P-DH}}{\rightleftharpoons}} \text{D-gluconate-6-P} + \text{NADPH} + \text{H}^+$$

6.6 Calibration of spectrophotometers

Periodic checks on the wavelength calibration and photometric accuracy of a spectrophotometer are essential for accurate work. Solid filters made from materials such as didymium have narrow bands with accurately known λ_{max} values and can be used to calibrate the wavelength scale. A solution of potassium dichromate in 0·005 M sulphuric acid has peaks at 257 and 350 nm, and troughs at 235 and 313 nm with accurately known absorbance values, and these can be used to check the photometric accuracy.

6.7 Sample presentation

UV–visible absorption studies are usually performed on samples in dilute solutions. In order to obtain accurate spectra, the solution must be optically clear and free from dispersed solid particles which would scatter the radiation rather than transmitting it. The solution is placed in a cell, or cuvette, which often has a 10 mm pathlength. Cuvettes with pathlengths between 1–100 mm are available. It may be desirable to analyse samples continuously from a system undergoing change, for example, the eluent from a chromatographic column, and this can be achieved with a flow cell through which the sample can be continuously pumped. For the visible region, glass or plastic cells may be used, but fused silica (quartz) or UV-transparent plastic cells are required for the UV region. The temperature of the cuvette should be controlled for accurate kinetic studies. This can be achieved with a cuvette block through which a thermostatically controlled liquid is circulated, or alternatively by means of electrical heating.

6.8 Difference spectrophotometry

Variations in the environment of a chromophore may produce small changes in a spectrum. Detection of small spectral changes can best be achieved by difference spectrophotometry which produces a direct plot of the difference in absorption of two samples in a double-beam spectrophotometer. One solution is placed in the reference beam and the other is placed in the sample beam. The difference spectrum may have both positive and negative peaks, as shown in Figure 6.8.

Difference spectrophotometry is used routinely by biochemists in studies of the changes in proteins due to pH, solvent or temperature. Changes in protein spectra with solvent can reveal which residues are exposed on the surface of a protein and hence subject to interactions with the solvent. Changes in protein conformation and denaturation can also be revealed by difference spectra. Dangerous drugs, such as barbiturates, may be analysed by difference spectrophotometry, since the spectra are sensitive to changes in pH which affect the ratio of keto to enol forms. Difference spectra are also used in the

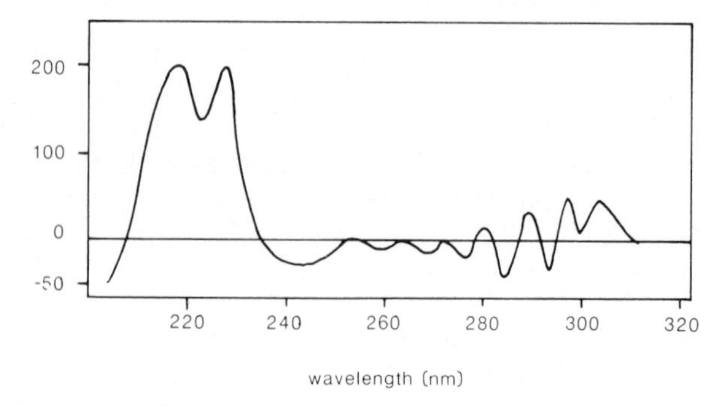

Figure 6.8 Solvent difference spectrum of phenylalanine produced by 20% (V/V) ethylene glycol/water compared with water. Redrawn with permission from Donovan (1969).

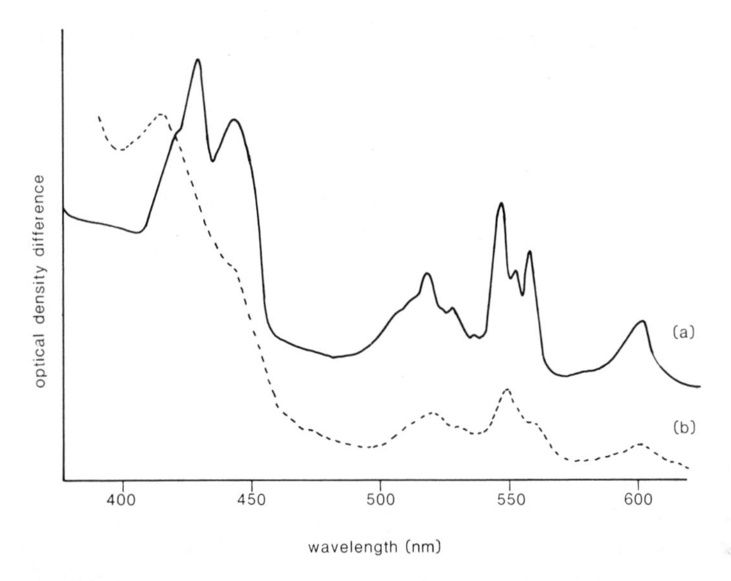

Figure 6.9 Difference spectra of baker's yeast: (*a*) low temperature, and (*b*) room temperature.

detection and quantification of electron carriers such as cytochromes (Figure 6.9) with the oxidized form in one beam and the reduced form in the other.

6.9 Spectrophotometric titrations

UV–visible spectrophotometry may be used to follow the course of a titration and to determine the end point of reactions in which one of the reactants or the product absorbs in a suitable part of the spectrum. The plot of absorbance against volume of titrant added allows the end point to be determined. The end point corresponds to the titre where the straight-line sections of the titration curve intersect, as shown in Figure 6.10. The gradients of the straight-line sections may be positive, negative or zero, depending on the absorption of the reactants and product at the wavelength being monitored. Spectrophotometric titrations are most sensitive if absorption of radiation by molecules not involved in the reaction is small compared with the change in absorption during the titration. Also, it is convenient if the absorbance remains on scale during the titration, obviating the necessity for dilution of samples. Selection of a wavelength other than λ_{max} may be beneficial in meeting these conditions. Some systems exhibit isobestic points. These are wavelengths at which the absorbance curves of the reactant and product molecules intersect. The molar absorbance is the same for both species at the isobestic points, and hence monitoring of a spectrophotometric titration must be performed at wavelengths well away from the isobestic points.

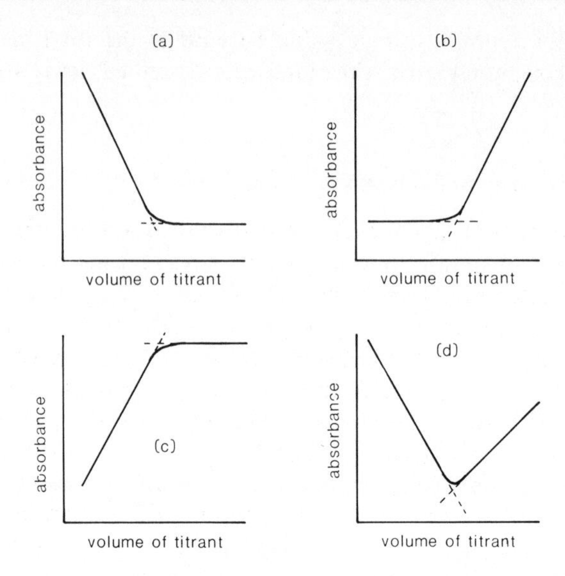

Figure 6.10 Titration curves. (*a*) Absorbing reactant converted into non-absorbing product; (*b*) titrant alone absorbs; (*c*) product alone absorbs; (*d*) absorbing reactant converted into non-absorbing product by an absorbing titrant.

Spectrophotometric titrations are particularly useful for monitoring reactions in which the absorption change occurs in the UV or the colour change is not distinct at the end point. The change in absorption of proteins with pH as alkali or acid is added allows the pK value of proteins containing weakly acidic groups to be determined if the ionization is reversible. This generally involves monitoring the absorbance arising from the phenolic groups of tyrosine by measurement at 295 nm. The absorbance shows a sharp change in the region of the pK values, as shown in Figure 6.11. The value of pK determined in this way

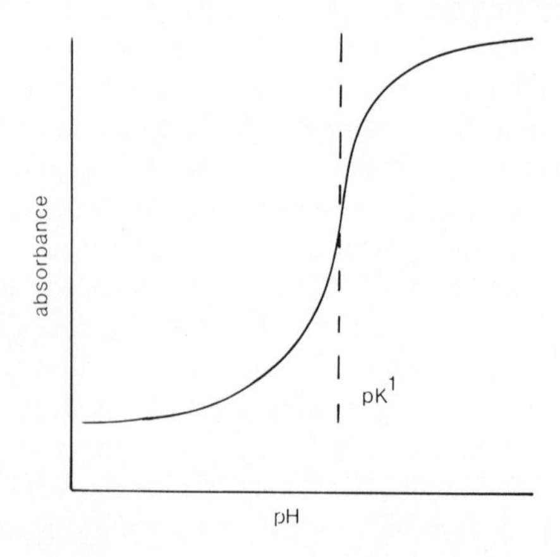

Figure 6.11 Schematic diagram of a typical spectrophotometric titration.

differs from the thermodynamic value because of the total charge on the protein. Spectrophotometric titrations are described in more detail by Headridge (1961).

6.10 Derivative spectrophotometry

Recording the first- or higher-order derivative spectrum is often beneficial in UV and visible absorption studies, since improved resolution of overlapping peaks and greater precision in determinations of λ_{max} may be achieved. Modern spectrophotometers are commonly constructed with a facility for direct recording of derivative spectra. A first-order derivative spectrum is the plot of the rate of change of absorbance A with wavelength λ, that is, $dA/d\lambda$ against λ. The second-order spectrum involves the use of the second derivative $d^2A/d\lambda^2$ as ordinate. The presentation of a symmetrical absorbance peak as its first- or second-order derivative spectrum is shown in Figure 6.12. The points of inflection in the original spectrum result in a maximum or minimum in the first derivative, while the absorption maximum becomes a zero crossing point in the first derivative. In the second derivative, the points of inflection of the original spectra appear at zero and the absorption maximum becomes a minimum. The vertical distance between a maximum and an adjoining minimum in a derivative spectrum is proportional to the concentration of the solute and this can be used for quantitative measurements.

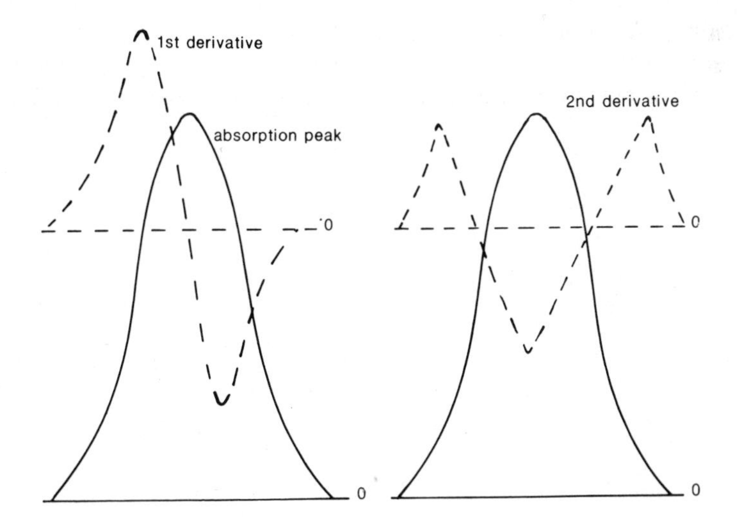

Figure 6.12 Symmetrical absorption peak and its first and second derivative (theoretical example).

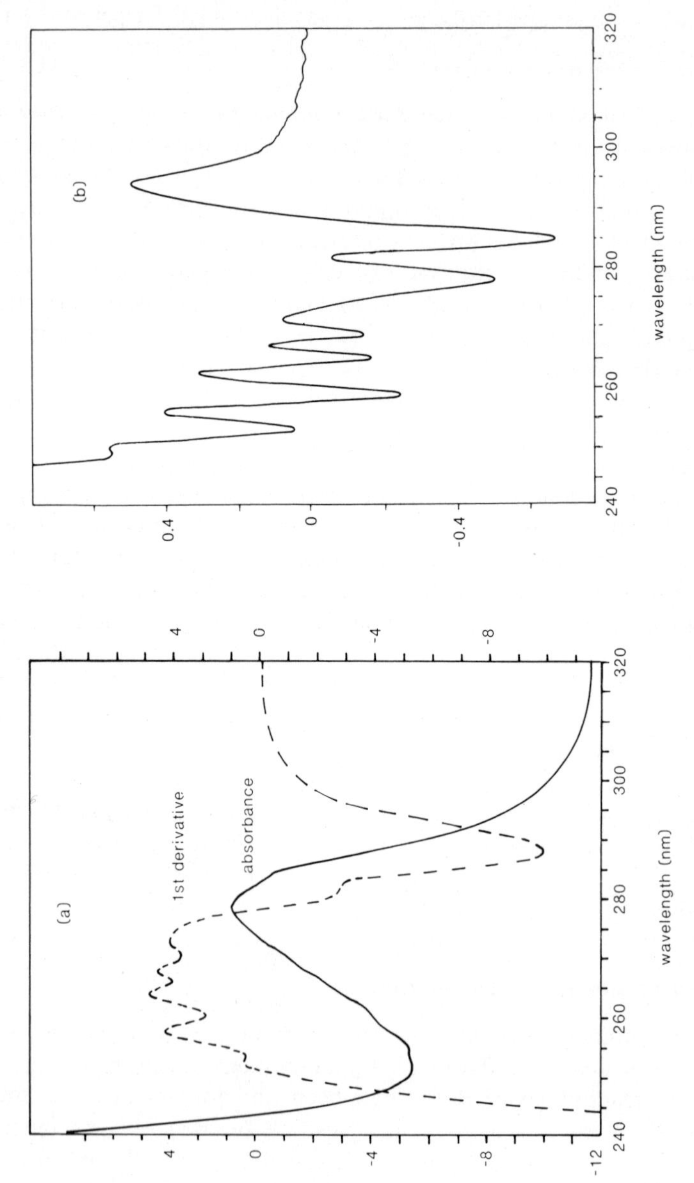

Figure 6.13 Ultraviolet spectra of bovine serum albumin (1 mg/ml). (*a*) Absorbance and first derivative; (*b*) second derivative spectra. (courtesy of Perkin–Elmer Ltd).

Derivative spectrophotometry is valuable for several types of determination. The first derivative can be used to measure the λ_{max} value of a broad peak, since the zero crossing point can be determined accurately. In practice, the resulting λ_{max} must be corrected by a wavelength shift dependent on the instrument settings. Derivative spectra show good resolution of overlapping absorption bands, and the second derivative spectrum is often used as a test for the purity of a material. The first- or second-derivative spectrum is also used for quantitative determinations in samples where an absorption band is superimposed on a wide band of a second component or broad background absorption due to turbidity. Figure 6.13 illustrates spectra obtained in the analysis of bovine serum albumin. The characteristic broad aromatic absorption band around 280 nm displays fine structure in the first derivative. The second derivative spectrum can be used to quantify the phenylalanine content from the bands at 250–270 nm, while the tyrosine and tryptophan residues absorb between 275 nm and 300 nm.

6.11 Dual-wavelength spectrophotometry

In a dual-wavelength spectrophotometer, radiation at two wavelengths passes through the sample alternately. If the difference in wavelength is small, typically 1 or 2 nm, the difference in absorbance values can be monitored as the spectrum is scanned and represented as the derivative spectrum. Alternatively, a spectrum can be obtained by scanning one wavelength while the other is fixed. This technique is very useful in the analysis of turbid samples where light scattering interferes with the absorption spectrum. Dual-wavelength spectrophotometry can also be used to eliminate the spectral contribution of an interfering component in a sample. This procedure requires the two wavelengths to be set at values at which the interfering component has identical molar absorption coefficients, while the analyte must absorb more strongly at one of these wavelengths than at the other. The absorbance changes at two fixed wavelengths can also be monitored simultaneously to determine changes in two components of the sample.

6.12 Spectrophotometers and colorimeters

A spectrophotometer suitable for studies in the UV and visible region comprises a source of radiation, a monochromator, a sample chamber, detector and readout device. Both single-beam and double-beam instruments are commonly used. A typical scanning double-beam spectrophotometer is shown in Figure 6.14. A single-beam instrument is generally only used for quantitative measurements at fixed wavelengths. If the instrument employs filters instead of a monochromator and is limited to the visible region, it is referred to as a colorimeter. Since colorimetry may also be performed using a colour comparator, which involves a visual comparison of samples, an

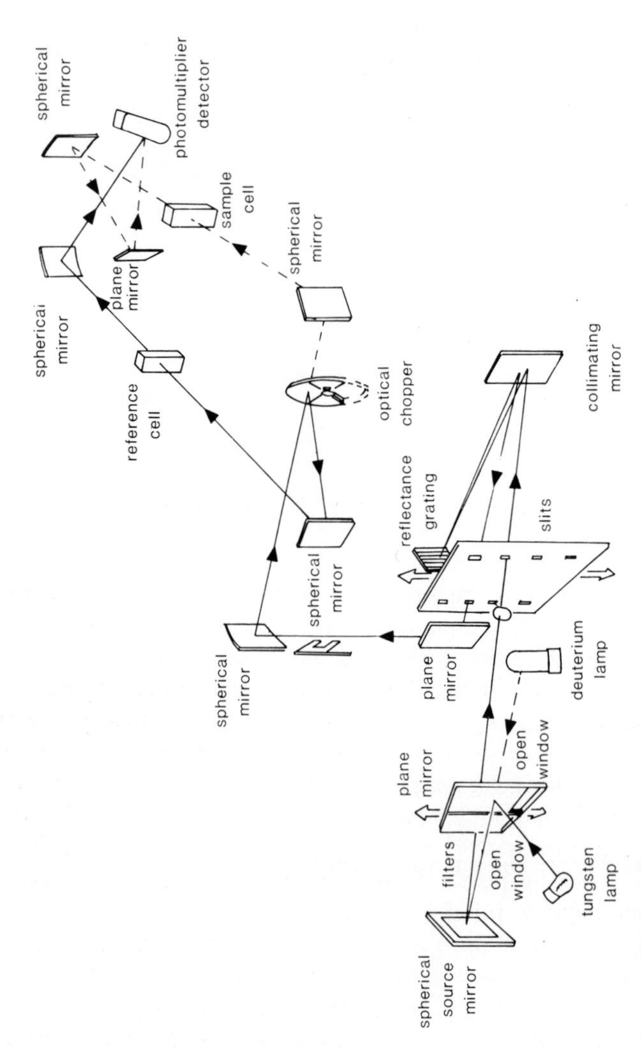

Figure 6.14 Optical system of UV–visible spectrophotometer (courtesy of Perkin–Elmer Ltd).

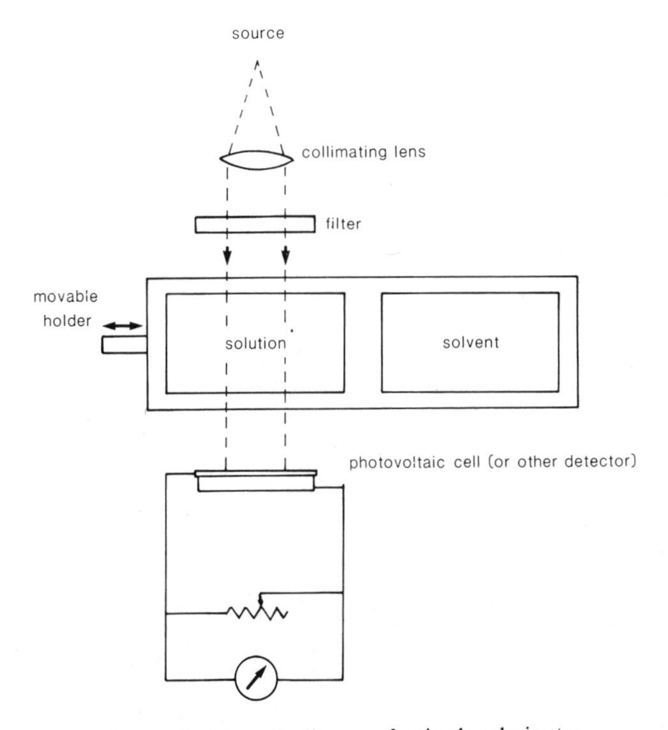

Figure 6.15 Schematic diagram of a simple colorimeter.

instrument fitted with a quantitative detector is called a photoelectric colorimeter or absorptiometer. Quantitative measurements with a single-beam instrument are performed by setting the absorbance to zero (100% transmittance) with a reference solution and then replacing this with an identical cuvette containing the sample solution and reading off the absorbance from a meter or numerical display. The key features of a simple colorimeter are shown in Figure 6.15. The filters used in a colorimeter transmit a broad band of radiation compared with a monochromator. The spectral bandwidth or bandpass of a monochromator or filter is the difference in wavelength between the points where the transmittance of the peak is half the maximum (see Figure 6.16).

6.12.1 *Radiation source*

The source must provide light of a suitable intensity and stability over a wide range of the spectrum. Spectrophotometers usually use a tungsten-filament lamp as a source for the spectrum from 330–850 nm, and a hydrogen- or deuterium-gas discharge lamp from 200–450 nm. The tungsten-filament lamp radiates as a black body when heated to about 2800 K. The filament is coiled and enclosed in a hermetically sealed bulb of glass that is evacuated or filled with an inert gas. The lamp is robust and relatively low-cost. Although the wavelength for maximum emission of radiation from a tungsten filament lies in the near IR at about 1000 nm, the lamp is sufficiently bright for use in the visible region. Good stability of the emitted radiation can be achieved by providing a constant voltage to heat the filament.

The hydrogen or deuterium lamp contains the gas at a low pressure (0·2–5 Torr). A low voltage (about 40 V direct current) is applied between the cathode and anode. The discharge produces a radiating ball of light which is restricted to a narrow path by a mechanical aperture between the electrodes. Deuterium provides a more intense beam of radiation compared to hydrogen and is the more common source for the UV region. The lamp must be pre-heated before use.

6.12.2 *Monochromators*

The monochromator converts the broad band of radiation emitted from the source to a narrow wavelength band. Early monochromators employed a prism but diffraction gratings are widely used in modern spectrophotometers. The refractive index (and hence angle of refraction) of a prism decreases with increasing wavelength of radiation and so the spectrum can be scanned by rotating the prism to allow radiation of specific wavelengths to be focused on the exit slit. Alternatively, a mirror mounted behind the prism can be rotated to produce the different wavelengths at the exit slit. Diffraction gratings have advantages of improved resolution, wider range and lower cost compared with prisms. A diffraction grating is a regular array of grooves on a surface with the pattern repeating at a distance d (Figure 6.17). Radiation of wavelength λ

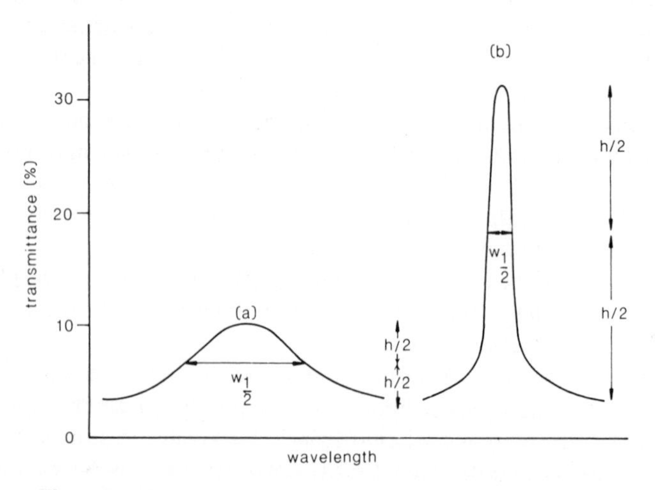

Figure 6.16 Bandpass, $w_{1/2}$, of (*a*) a filter or (*b*) a monochromator.

striking the grating at an angle i to the normal is reflected at an angle r with constructive interference if

$$m\lambda = d(\sin i + \sin r)$$

Radiation which is not at the wavelength λ is destroyed by destructive interference. Several orders of spectra, corresponding to the reflection of light in certain directions, arise from the fact that m can have small integer values. The wastage of energy in unwanted orders is reduced by the use of sawtooth-shaped grooves in the grating. The grating is then described as blazed, with a concentration of reflected radiation in a given direction. For light at the blaze wavelength λ_b, where

$$m\lambda_b = d \sin \alpha$$

the grating acts as a mirror and reflects nearly all the radiation. In practice, a grating is useful over a restricted range between $\frac{2}{3} - \frac{1}{2}$ of the blaze wavelength. The resolution of a diffraction grating is equal to mN, where N is the number of lines ruled on the grating and m is the order of the reflected light. Gratings for use in the UV–visible range may be 50 mm wide with 600 grooves mm^{-1}. This yields a resolving power at 300 nm of 0·01 nm. Modern gratings are generally plastic replicas of an original ruling on a metal surface.

6.12.3 *Filters*

Filters may be used to produce a narrow band of radiation in instruments where only a limited number of fixed wavelengths are required. They are also used in colorimeters where a wide band of incident radiation is acceptable. Narrow bandwidths are provided by interference filters which consist of two parallel, partially reflecting silver films separated by a thin dielectric spacer film (CaF_2, MgF_2 or SiO_2). Constructive interference can occur between a beam of light transmitted directly through the film and a beam transmitted after reflection within the spacer film.

Absorption filters are commonly made of coloured glass for use in colorimeters, but other materials including gelatin or plastic may also be used.

6.12.4 *Detectors*

The most common detector for scanning spectrophotometers is the photomultiplier, but other detectors, including photovoltaic cells and photodiodes, may also be used for special purposes.

In a photomultiplier (Figure 6.18) the incident light falls on a photosensitive surface, often containing caesium, and each photon causes the release of an electron. An electric potential of 70–100 V accelerates this towards a curved electrode, called a dynode, coated with a compound such as beryllium oxide or gallium phosphide, whose surface liberates several secondary electrons as a

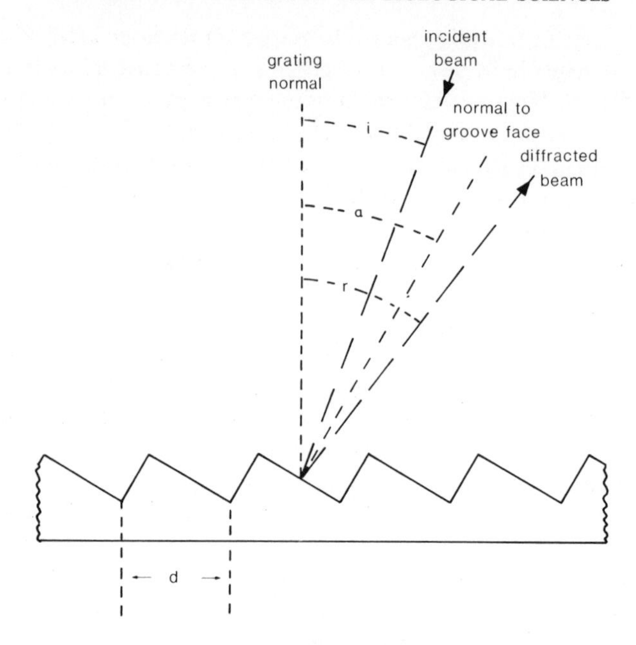

Figure 6.17 A diffraction grating.

result of the impact of a high-energy electron. Each secondary electron is accelerated electrically to another dynode at a higher potential, where again each electron releases several secondary electrons. This electron multiplication process is repeated for 10 to 15 stages and the total current amplification can reach several million. Even in the absence of incident light, a small current still flows and is amplified. This dark current must be minimized in the design of the photomultiplier, but since it represents a steady component, it can be subtracted automatically from the signal, as the instrument is zeroed.

Colorimeters may use a photovoltaic cell as the detector. This consists of a thin layer of a semiconductor, such as selenium, coated onto a metal base plate. A very thin layer of silver or gold is sputtered over the selenium. Incident light falling on the selenium produces electron–hole pairs at the noble metal–selenium interface. A small potential develops between the base plate and the noble-metal surface and a current flows in the circuit which is designed with a low resistance. The current is approximately proportional to the intensity of the incident light. Because the cell impedance is low, the current cannot be amplified with a simple amplifier and therefore this detector is mainly used in filter colorimeters, which provide a high level of incident light allowing the current to be monitored directly without amplification.

The importance of UV detection in HPLC has led to the development of computer-aided rapid-scanning detectors, which reduce the time required for measuring a full wavelength spectrum from several minutes to one second. The most important of these at present is the linear photodiode array (Cahill and Retzik 1985). White light passes through the sample and a diffraction grating partitions the spectrum into several hundred wavelength components, each of which falls on a separate diode in a linear array which is typically about 1 cm long. The light energy in each of the wavelength components is simultaneously measured and then read out sequentially over a common output line to yield a conventional spectrum.

6.13 Turbidimetry and nephelometry

Samples which are not optically clear due to the presence of suspended particles cause light to be scattered as well as absorbed and transmitted. Turbidimetry involves the measurement of the intensity of the transmitted light as a function of the concentration of the dispersed phase and nephelometry involves the measurement of the intensity of the light scattered at right angles to the incident light beam as a function of the dispersed-phase concentration. The scattering of light is elastic, so that the wavelength of the scattered radiation is identical to that of the incident beam. Light scattering is dependent on various factors besides the concentration of the dispersed phase, including particle-size distribution, particle shape, temperature, refractive index and molecular absorption, and consequently it is necessary to construct a calibration curve with a number of standards. Generally, turbid solutions do not obey Beer's law.

Visual and photoelectric colorimeters may be used as turbidimeters, with a blue filter usually giving the greatest sensitivity. Nephelometry is generally more sensitive than turbidimetry for dilute suspensions. Low-cost nephelometers can be purchased, but most fluorometers may also be adapted for use in nephelometry, since the incident light beam is at right angles to the plane of the detector. The nephelometer shown in Figure 6.19 incorporates a 6 V, 6 W lamp in the base of the unit. Light shines vertically through the orifice of an annular photocell onto the hemispherical base of a test-tube. A wheel equipped with three colour filters and a white-light position is situated between the lamp and the photocell. The selected filter should be similar in colour to that of the sample. Light scattered by dispersed particles is collected by a reflector mounted above the photocell and directed onto the photocell. A metal cap fitting over the sample tube excludes extraneous light. The current flowing in the measuring circuit is measured by a sensitive galvanometer.

Nephelometry and turbidimetry are applied in a variety of fields. These include the determination of suspended matter in water and determination of the clarity of beverages and pharmaceutical preparations. Both techniques are commonly used in the measurement of bacterial concentrations (Koch 1981). Most of the light scattered from particles in the size range of bacteria is within 2–12° of the transmitted beam and a well-collimated incident light beam such as a laser beam is required for measurements within this range. Measurements

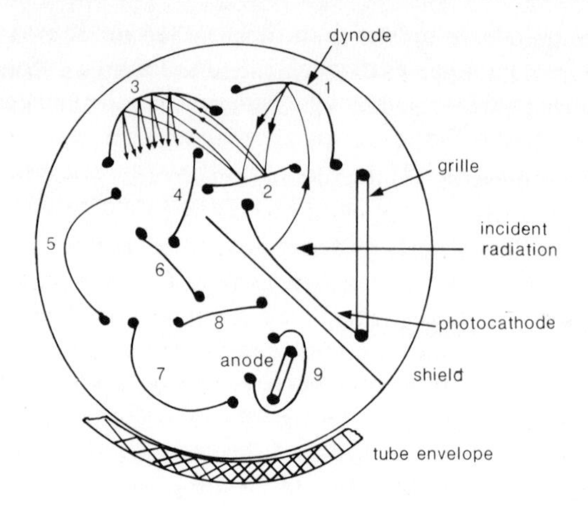

Figure 6.18 Photomultiplier detector.

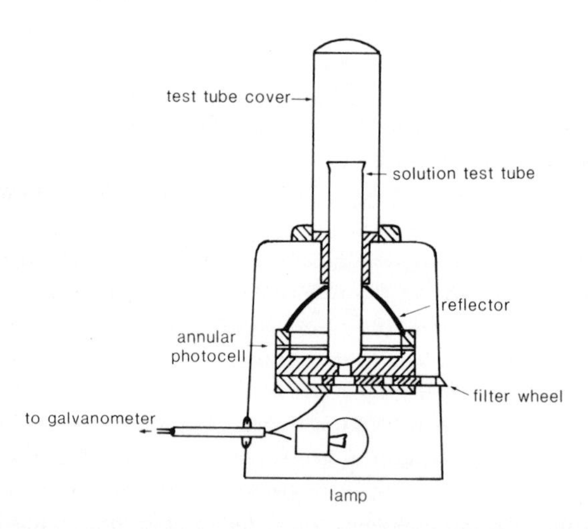

Figure 6.19 A commercial nephelometer (courtesy of CIBA-Corning Diagnostics Ltd).

at higher angles contain information about the amount of cell material, but they are also affected by the internal structure and distribution of material within the cell.

6.14 Colour and gloss of solid samples

The surface appearance of a solid material is dependent on the wavelengths of light absorbed by the surface, together with the angular distribution of the reflected light. These two properties determine the colour and gloss of the sample. White, grey and black surfaces reflect light equally at all wavelengths in the visible region. However, the proportion of the incident light reflected varies, being greatest for white surfaces and least for black surfaces. Coloured substances absorb light of certain wavelengths, but reflect light at other wavelengths.

The colour of a solid sample can be determined by matching the colour with that of a standard. Several methods have been developed for colour measurement. The CIE system considers colour to be defined in terms of three parameters namely the observer, the illuminant and the sample reflectance. In the early stages of colour measurement, red, green and blue were investigated as primary colours, but no simple combination of these colours can match all observed colours. Instead three imaginary primaries X, Y and Z, known as tristimulus values, were used. The tristimulus values allow for positive and negative contributions of primary colours in matching the colour of a sample. A negative contribution involves the addition of a primary to the sample to match a reference, whereas a positive contribution involves the addition of a primary to a reference to match a sample. In order to allow for the response of the eye to different colours, a large number of observers with normal colour vision were asked to match each colour of the spectrum by combinations of red, green and blue. A mean specification in terms of the amount of each primary required to match each spectral colour was obtained, and the standard observer curves were constructed to represent the mean response of the eye to primary-colour combinations. The CIE system has specified several standard illuminants which may be used for colour measurement. The colour of a sample can then be determined by matching it with a standard and defining the colour in terms of the tristimulus values, taking into account the standard observer curves, the type of illuminant and the sample reflectance.

The CIE system has significant limitations, in that the imaginary primaries have no direct relationship to the perceived colour, and the colour space defined by X, Y and Z is non-uniform, with small differences in the colour of blue samples being represented by small changes in X, Y and Z, but small differences in the colour of green samples being represented by larger differences in the tristimulus values. In order to overcome these limitations, other systems of colour measurement have been developed. The Hunter L, a, b system is widely used for food colorimetry. This system represents the colour of samples in terms of three parameters. The parameters a and b represent red to green, and blue to yellow colour dimensions. The third parameter, L, represents the lightness of the sample varying from $L = 0$ for black samples to $L = 100$ for white samples. The colour of a sample can be described by a combination of these three colour parameters.

6.14.1 *Gloss measurement.* The gloss of a sample represents the extent to which it reflects light, like a perfect mirror with the angle of reflection equal to the angle of incidence. Light that is reflected in this way is termed specular reflection. A matt surface has a high degree of scattered light and a small amount of specular reflection, while the reflection from a glossy surface is mainly specular. The gloss of pharmaceutical or food samples is often an important quality parameter. It may be measured using a gloss meter set at a single angle of incidence, such as 60°, or several angles of incidence may be used. A goniophotometer measures the reflection of light with a variable angle of incidence.

References

Cahill, J. and Retzik, M. (1985) *Int. Lab.* May, 48.

Donovan, J.W. (1969) *J. Biol. Chem.* **244**, 1961.

Goodwin, T.W. and Mercer, E.I. (1983) *Introduction to Plant Biochemistry*, 2nd edn, Pergamon, Oxford.

Headridge, J.B. (1961) *Photometric Titrations*, Pergamon, New York.

IUPAC (1979) *Standard Methods for the Analysis of Oils, Fats and Derivatives*, 6th edn, Pergamon, Oxford.

Koch, A.L. (1981) Growth measurement, ch. 11, in *Manual of Methods for General Bacteriology*, Gerhardt, P. *et al.* (eds), American Society for Microbiology, Washington.

Further reading

Grum, F. (1972) ch. 3, in *Physical Methods of Chemistry*, vol 1, pt III B, Weissberger, A. and Rossiter, B.W. (eds), Wiley-Interscience, New York.

Morton, R.A. (1975) *Biochemical Spectroscopy*, vols. 1 and 2, Adam Hilger, London.

Olsen, E.D. (1975) *Modern Optical Methods of Analysis*, McGraw-Hill, New York.

Rao, C.N.R. (1975) *Ultraviolet and Visible Spectroscopy* (*Chemical Applications*), Butterworth, London.

7 Fluorescence and phosphorescence spectrophotometry

7.1 Introduction

Luminescence is the emission of photons from electronically excited states. Fluorescence and phosphorescence are two types of luminescence which differ in the nature of the ground and excited states. A molecule may be excited from a singlet ground state to a singlet excited state by the absorption of radiation in the UV or visible region of the spectrum. Electronic excitation is accompanied by excitation to a higher vibrational energy level. The excited molecules may rapidly lose vibrational energy by collisions with neighbouring molecules. Fluorescence is the emission of light at a longer wavelength than the incident

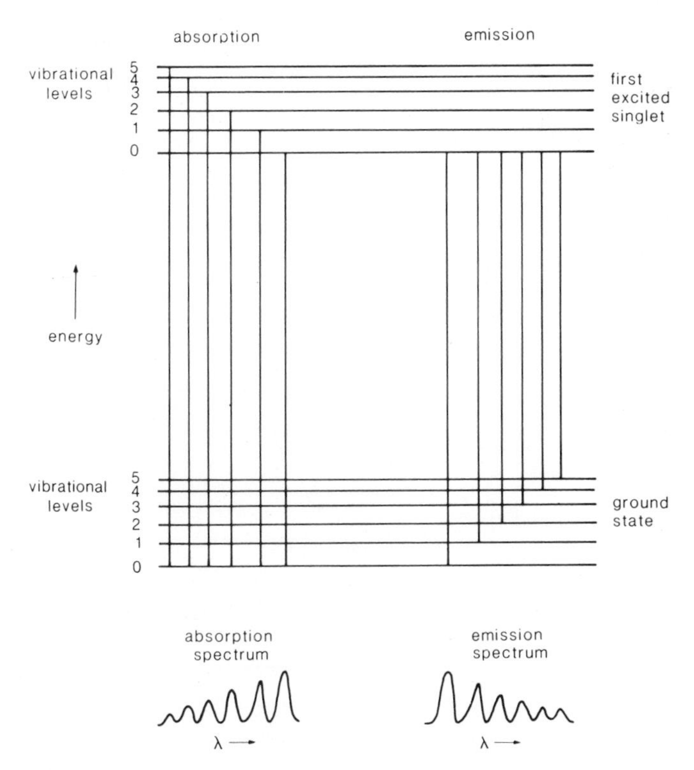

Figure 7.1 Energy level diagram and spectra for transitions between the ground and first excited singlet states.

radiation as the molecule returns from a singlet excited state to a singlet ground state after the loss of vibrational energy (Figure 7.1). Phosphorescence is the emission of light at a longer wavelength than the incident radiation as a molecule returns from an excited electronic state to a ground electronic state of different spin multiplicity, generally a triplet excited state returning to a singlet ground state.

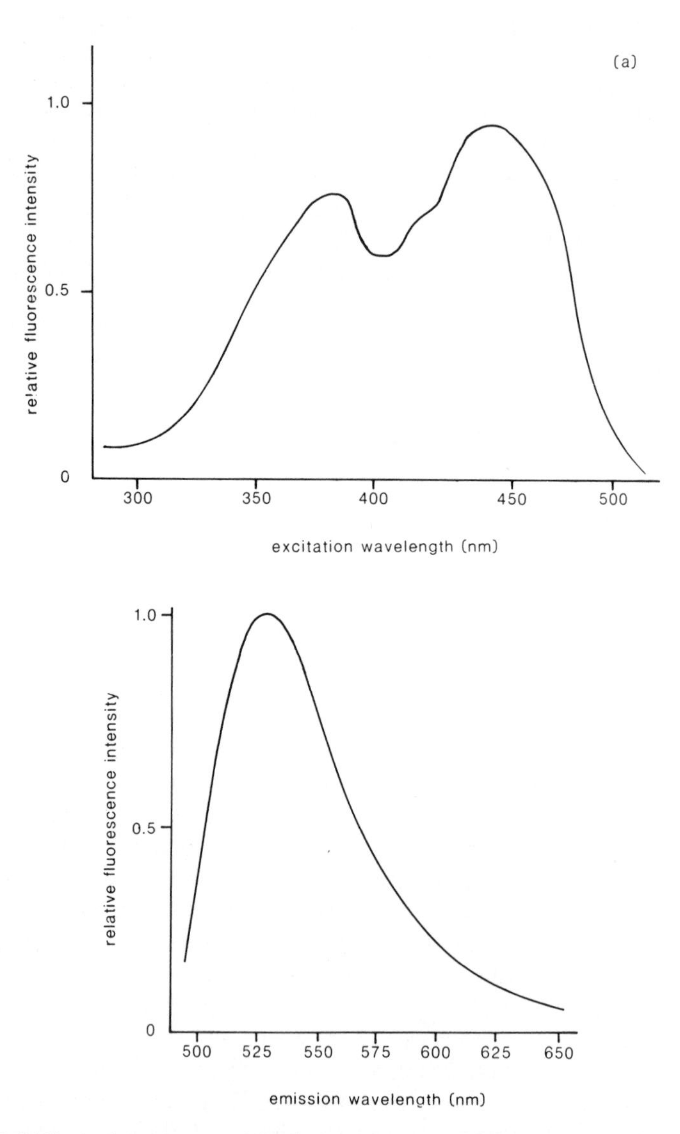

Figure 7.2 (*a*) Excitation spectrum of riboflavin (emission at 525 nm). (*b*) Emission spectrum of riboflavin (excitation at 480 nm).

Fluorescence is an allowed transition and therefore the fluorescence lifetime is short, typically about 10^{-8} s. Phosphorescence, however, is a forbidden transition and the emission rates are slow. Phosphorescent lifetimes range from 10^{-3}–10 s, depending primarily on the importance of deactivation processes other than emission. Fluorescence is observed at moderate temperatures in solution, while phosphorescence usually occurs in rigid samples at low temperatures, since non-radiative deactivation processes are usually faster than phosphorescence in fluids and at ambient temperatures.

Fluorescence spectrophotometry has become a very important analytical procedure in recent years. Although it is by no means universally applicable, it is often more sensitive and more specific than UV or visible spectrophotometry for detection of molecules that exhibit fluorescence. High sensitivity results from the fact that the emitted radiation is at a different wavelength from the exciting radiation. Thus the background signal is close to zero and this allows the signal to be detected more precisely than in procedures that involve the measurement of a small signal superimposed on a large background. Fluorescent materials are characterized by an excitation and an emission spectrum as shown for riboflavin (Figure 7.2). This gives the analytical procedure a good degree of specificity, since molecules with similar absorption spectra may differ in their fluorescence spectra. Fluorescence is a common means of detection for HPLC (see Section 2.4.4).

7.2 Fluorophores

Molecular structures which fluoresce are termed fluorophores. Fluorescent molecules are typically aromatic or molecules with multiple conjugated double bonds. Molecules with these structures possess low-lying electronic excited states to which excitation is possible with radiation wavelengths greater than 220 nm, that is, energies less than 550 kJ mol^{-1}.

Not all molecules that absorb radiation in the visible or UV region exhibit fluorescence. Non-radiative processes may take place, resulting in the loss of energy from the excited state, and fluorescence is not observed if these processes are fast compared with the emission of radiation. Radiationless transitions between a singlet excited state and the singlet ground state may occur if the potential-energy diagrams intersect, that is, if the energy of the two electronic states is identical at a particular internuclear separation (Figure 7.3). If the exciting radiation generates molecules in the vibrational energy level corresponding to this energy, a radiationless transition to the ground state occurs. This is known as internal conversion.

Nucleotides, lipids, carbohydrates and non-aromatic amino acids are among the important classes of compounds that do not fluoresce, but some drugs and vitamins exhibit strong fluorescence. Fluorescence generally occurs if the transition to the lowest excited state is $\pi \rightarrow \pi^*$, while molecules which have a low energy $n \rightarrow \pi^*$ transition undergo intersystem crossing from the

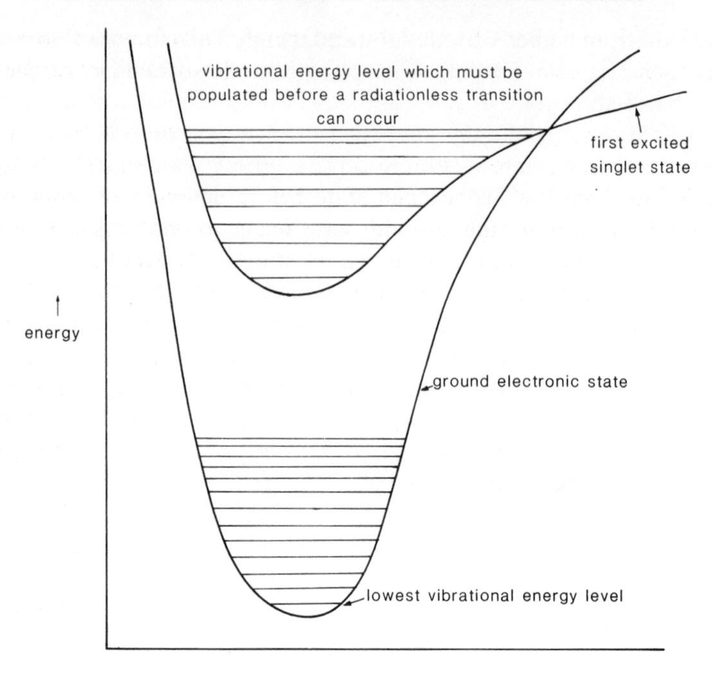

Figure 7.3 Effect of internuclear separation on the energy of two electronic states with the conditions for radiationless transitions.

singlet to the triplet state followed by phosphorescence or non-radiative transitions.

7.3 Excitation and emission spectra

Both excitation and emission spectra are useful for characterizing fluorescent compounds. The spectra can be recorded with a spectrofluorometer by first varying the excitation wavelength until fluorescence occurs; this can often be observed visually. The excitation wavelength is then fixed at this value and the emission spectrum is recorded by measuring the fluorescent radiation while scanning the emission wavelength. The emission wavelength is then set to the value corresponding to the maximum fluorescence and the excitation spectrum is recorded by measuring the fluorescent radiation as the excitation wavelength is scanned.

The excitation spectrum of a fluorescent molecule is often very similar to the absorption spectrum, although the former involves monitoring the emission of radiation from excited species in the lowest vibrational energy level of the first excited state and the latter relates to the absorption of radiation during excitation to this and other energy levels. The efficiency with which vibrational energy is lost from higher vibrational states is high. Therefore the fluorescence intensity at a particular excitation wavelength is often dependent on the total energy absorbed.

The fluorescence emission spectrum of a molecule occurs at longer wavelengths than the absorption spectrum, primarily because vibrational energy is lost from the first excited state before emission of radiation. In addition, the return to the ground state leads to molecules in various vibrational energy levels, as is shown in Figure 7.1. Another effect also contributes to the wavelength difference. According to the Franck–Condon principle, the absorption of radiation is fast and no movement of atomic nuclei occurs during this process (Figure 7.4). However, absorption of radiation frequently leads to a change of electron distribution and polarity in a molecule. Therefore solvent molecules are in non-equilibrium positions after absorption. Before radiation is emitted from the excited state, the atoms of the molecule relax to a configuration of lower energy, and the surrounding solvent molecules can be realigned to minimize the energy of the system. Emission of radiation is also fast and therefore the ground state generated may be one in which the solvent molecules are not in their lowest energy configuration. Consequently, the emitted radiation is lower in energy than the energy

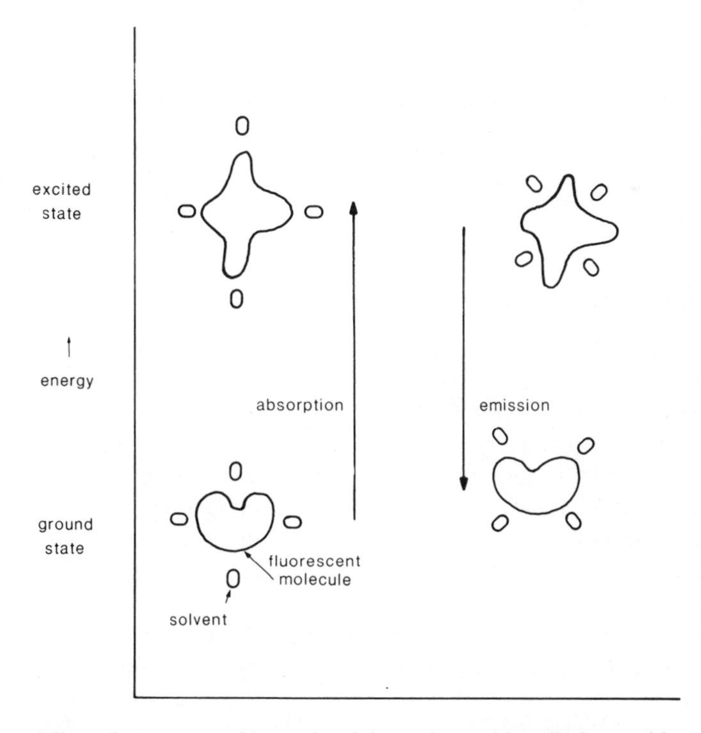

Figure 7.4 Effects of solvation on the energies of electronic transitions. Redrawn with permission from Brown (1980).

difference between the appropriate states at equilibrium and the absorbed radiation is correspondingly higher.

The emission spectrum often has the appearance of the mirror image of the absorption spectrum. This occurs because if the most probable absorption is from the zero vibrational energy level of the ground state to the *n*th vibrational energy level of the excited state, the most probable emission is often from the zero vibrational energy level of the excited state to the *n*th vibrational energy level of the ground state. This arises from the Franck–Condon principle for molecules in which there is a similar spacing between vibrational energy levels in the ground and excited states (Figure 7.1). There are, however, many samples that do not obey the mirror-image rule. Deviations are often observed

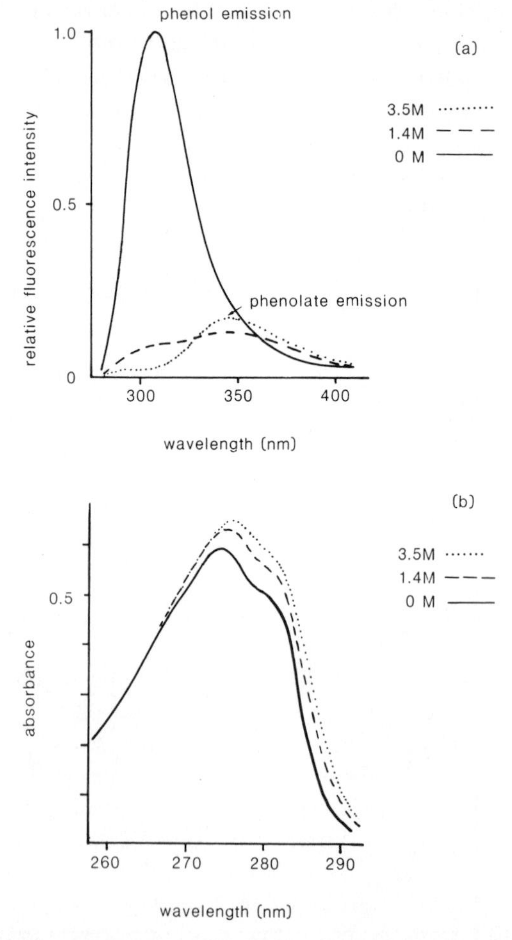

Figure 7.5 Effect of three concentrations of K_2HPO_4 on (*a*) the fluorescence spectrum, and (*b*) the absorption spectrum of tyrosine at pH 7.5. Redrawn with permission from Shimizu and Imakuro (1977).

for molecules which have a different geometric arrangement of nuclei in the excited state from that in the ground state, since this corresponds to different spacing of the vibrational energy levels in the two states. Reactions of the excited state may occur prior to emission of fluorescence and these also lead to differences in the absorption and emission spectra. An example of a reaction of an excited state is shown by tyrosine in the presence of a proton acceptor, which removes the phenolic proton following excitation. Depending on the concentration of the proton acceptor, either the phenol or the phenolate emission may dominate the emission spectrum (Figure 7.5).

7.4 Quantitative measurements

Fluorescence is linearly proportional to concentration at low concentrations in simple systems containing a single fluorophore in the absence of quenching effects. However, calibration curves should always be determined for accurate measurements. The Beer–Lambert law cannot be applied directly, since it relates to the intensity of radiation transmitted by a solution and fluorescence involves the emission of radiation. However, the relationship between concentration and fluorescence can readily be derived.

The fluorescence process can be represented by:

$$F \xrightarrow{k_1 I_0} F^* \begin{cases} \xrightarrow{k_2} F + h\nu & \text{(fluorescence)} \\ \xrightarrow{k_3} F & \text{(non-radiative quenching)} \end{cases}$$

where F is the ground-state fluorescent molecule, F* is the first excited singlet, I_0 is the intensity of exciting radiation and k_1, k_2, k_3 are rate constants. At a steady-state concentration of F*, $k_1 I_0[\mathrm{F}] = (k_2 + k_3)[\mathrm{F}^*]$. Fluorescence emission intensity is equal to $k_2[\mathrm{F}^*]$. Elimination of [F*] from these expressions gives

$$\text{Fluorescence intensity} = \frac{k_1 k_2 I_0[\mathrm{F}]}{k_2 + k_3} \qquad (7.1)$$

At low concentrations, [F] approximates to the total concentration of fluorescent molecules and therefore the fluorescence intensity is linearly proportional to this concentration. The constant of proportionality is dependent on the dimensions of the incident light beam, the area of solution irradiated, the luminescence efficiency or quantum yield, that is, the ratio of quanta emitted to quanta absorbed, the fraction of the fluorescence incident on the detector and the detector efficiency.

The above derivation shows that the intensity of the emitted radiation is proportional to concentration at low concentrations. However, a negative deviation from linearity usually occurs at higher concentrations and sometimes the fluorescence intensity will even fall with increased concentration at high levels. Thus, for the coenzyme NADH in aqueous solution, the fluorescence is linearly proportional to concentration between 10^{-8} and 10^{-4} M, but the fluorescence is reduced at higher concentrations. This occurs partly because the total concentration of fluorescent molecules is different from the ground-state concentration and the correct value should be used in equation (7.1).

Another effect contributing to the deviation from linearity is that the excitation and emission optics of the fluorometer focus on the centre of the sample cuvette. Increased concentration leads to increased absorption of incident radiation before it reaches the centre of the sample cuvette and also increased absorption of fluorescence by the sample, and therefore an inner filter effect reduces the effective intensity of the emitted radiation. Self-quenching effects in which radiation is lost by non-radiative transitions due to the collision of two molecules in their excited states may also contribute to the deviation from linearity.

7.5 Factors affecting fluorescence spectra

7.5.1 *Quenching of fluorescence*

Fluorescence quenching refers to any process which decreases the fluorescence intensity of a substance. These processes include collisional quenching, complex formation, energy transfer and excited-state reactions.

Collisional quenching occurs when a quencher molecule collides with the fluorophore causing it to return to the ground state without emission of a photon. Collisional quenchers include molecular oxygen, chloroform and some metal ions. The quencher must diffuse to the fluorophore during the lifetime of the excited state. Therefore, studies of collisional quenching can be used to determine the diffusion rate of quenchers and information can also be gained about the location and accessibility of fluorophores in proteins and membranes. An example of this type of study was the investigation of the quenching of the fluorescence of tryptophan residues in the protein trypsinogen by oxygen and iodide ions (Lakowicz and Weber 1973). Oxygen gives considerable quenching with a quenching constant of about 40% of that expected for a diffusion-controlled reaction. However, very little quenching was observed for iodide ions. It was deduced that most of the tryptophan residues were in the interior region of the protein and that charged and hydrated iodide ions cannot penetrate the non-polar interior of the protein.

An alternative quenching mechanism is complex formation, or static quenching, which involves the formation of a non-fluorescent ground-state complex between the fluorescent molecule and the quencher, causing the system to return to the ground state without the emission of a photon.

There are several methods of distinguishing static and collisional quenching. Collisional quenching reduces the lifetime of the excited state by increasing the efficiency of deactivation processes, while static quenching does not affect the excited-state lifetime, but reduces the concentration of free fluorophore available for excitation. Increasing temperature leads to an increase in collisional quenching, but a reduction in static quenching. In addition, static quenching may affect the absorption spectrum of the fluorophore, but collisional quenching does not. However, the change in the absorption spectrum is not always evident.

Fluorescence energy transfer involves the transfer of excited-state energy between a donor molecule and an acceptor molecule without the appearance of a photon. This quenching mechanism arises from dipole–dipole interactions between the two molecules. The rate of energy transfer depends on several factors, including the extent of overlap of the donor emission spectrum with the absorption spectrum of the acceptor molecule, the relative orientation of the dipoles of the two molecules and the distance between them. Energy transfer affects the spectra of macromolecules containing several fluorophores. For example, the emission spectra of most proteins are dominated by tryptophan in spite of the presence of tyrosine and phenylalanine which exhibit fluorescence as amino acids (Figure 7.6). This occurs because tryptophan absorption and emission are the lowest energy transitions of the three amino-acid residues and, in most proteins, tyrosine and phenylalanine transfer energy from their excited states to tryptophan more rapidly than emission of a photon. Other effects including quenching by carboxy or amino groups also play a role in protein fluorescence.

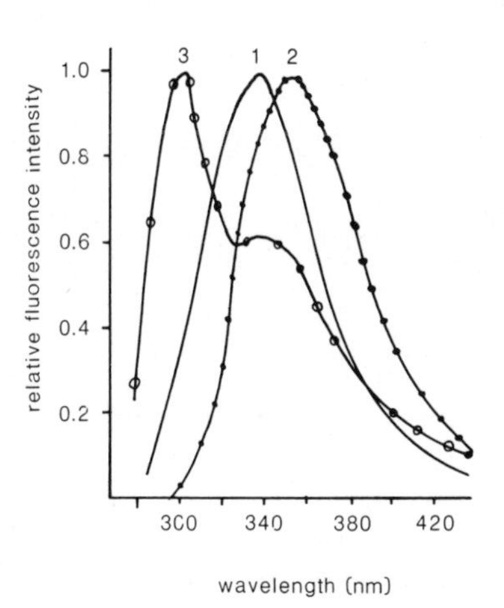

Figure 7.6 Fluorescence emission spectra of (1) human serum albumin, (2) tryptophan, (3) a mixture of tyrosine and tryptophan equal to that in human serum albumin. (Excitation at 240–250 nm.) Redrawn with permission from Weinryb and Steiner (1971) and Vladimirov and Burstein (1960).

7.5.2 *Solvent effects*

As discussed in Section 7.3, solvent molecules can affect the energy of electronic transitions, and therefore large solvent effects on the excitation and emission spectra can be observed. The excited state is usually more polar than the ground state and it is generally found that polar solvents shift the fluorescence emission to longer wavelengths. Solvents may also affect the intensity of fluorescence, that is, the quantum yield. This may occur as a result of various types of specific interactions. For example, many molecules containing oxygen or nitrogen, including aromatic alcohols and heterocyclic nitrogen compounds, show increased fluorescence in hydroxylic solvents compared to hydrocarbons. In hydrocarbon solutions, the lowest energy transition is $n \rightarrow \pi^*$, with the excited electron located on oxygen or nitrogen in the ground state. This excitation leads to triplet-state formation rather than fluorescence. However, in hydroxylic solvents, the lone pair is involved in hydrogen bonding to the solvent and the $\pi \rightarrow \pi^*$ transition becomes lowest in energy. This leads to increased fluorescence. Quenching may also occur with some solvents, particularly if the solvent molecule contains a large polarizable atom such as iodine.

Solvent effects on fluorophores bound to macromolecules and membranes are often used to determine the polarity and nature of the environment close to the binding site. Waggoner and Stryer (1970) showed that the fluorescence of 12-(9-anthroyloxy)-stearic acid bound to phosphatidylcholine bilayers was similar to that in solutions of hexane and benzene (Figure 7.7). They therefore concluded that the molecule was bound to the non-polar fatty-acid region of the bilayer and could be used to monitor changes in this region.

7.5.3 *The effect of pH*

The fluorescence spectra of the protonated and unprotonated forms of a molecule may be quite different. The wavelength may be shifted or only one form may fluoresce at a particular wavelength. If only one form fluoresces, the

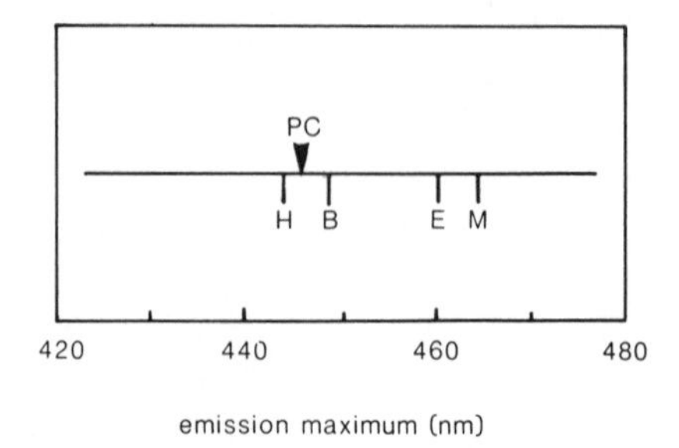

Figure 7.7 Emission maximum of 12-(9-anthroyloxy)-stearic acid bound to phosphatidylcholine bilayers (PC) and in hexane (H), benzene (B), ethanol (E) and methanol (M). Redrawn with permission from Waggoner and Stryer (1970).

molecule may be used as an acid–base indicator, with the pK_a defined as the pH value at which the fluorescent intensity is half the maximum value. The excitation pK_a and the emission pK_a are not identical. Thus for tyrosine, the pK_a of the aromatic hydroxyl group is 10 in the ground state, but falls to 4 in the excited state.

7.5.4 *Polarization effects*

If a fluorophore is excited with plane-polarized light, that is, light for which the component electric field oscillates in a single plane, the emitted radiation is often partially depolarized. This occurs because the axes for absorption and emission are not necessarily coincident due to differences in structure between the ground and excited states. In addition, the molecule may have time to rotate between absorption and emission of radiation. This dependence of depolarization on molecular rotation has been used to investigate the mobility of macromolecules and the micro-viscosity of cell membranes to which a fluorophore has been bound. Protein denaturation and protein–ligand association can also be investigated by this technique.

Fluorescence polarization has also been applied in immunoassays. When fluorescent-labelled antigens are bound to a large antibody molecule, they form a massive molecule that tumbles more slowly and emits light polarized in the same plane as the incident light. Thus the degree of depolarization falls as the amount of antibody binding increases.

7.5.5 *Fluorescence lifetimes*

The fluorescence lifetime τ of a substance is the average period of time that a molecule spends in the excited state prior to its return to the ground state. A commonly used method of studying fluorescence lifetimes involves the

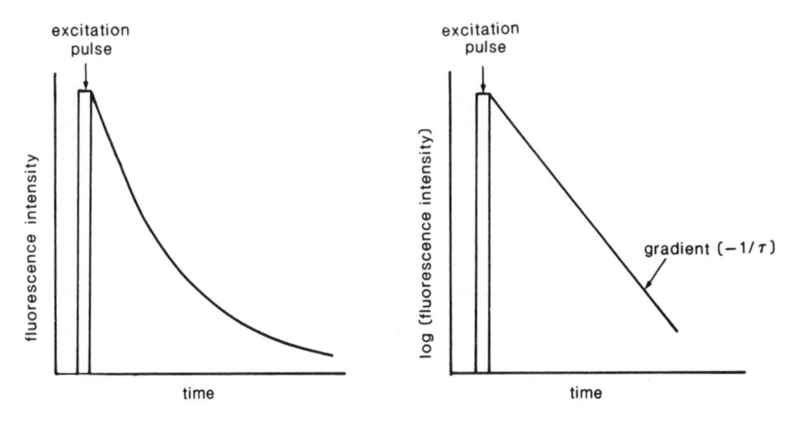

Figure 7.8 Observed decay of fluorescence following a pulse of exciting radiation.

illumination of a sample with a brief pulse of light followed by the determination of the decay of fluorescence intensity. For a simple system, decay follows a first-order rate law. The exponential decay observed (Figure 7.8) can be plotted as a semi-log plot, with the gradient $-1/\tau$. The lifetime of the excited state is dependent both on the rate of fluorescence emission and the rate of non-radiative decay.

The measurement of fluorescence lifetimes requires more complex equipment than constant illumination spectrofluorometers. However, useful information about the interactions of the fluorophore with its environment, rates of molecular motion and excited-state properties can be obtained from these investigations.

7.6 Instruments for fluorescence studies

Instruments for the measurement of fluorescence are known as fluorometers, fluorimeters, spectrofluorometers or spectrofluorimeters. The basic components of a fluorometer are:

(a) a source of radiation for excitation of the fluorophore
(b) a filter or monochromator to select the excitation wavelength

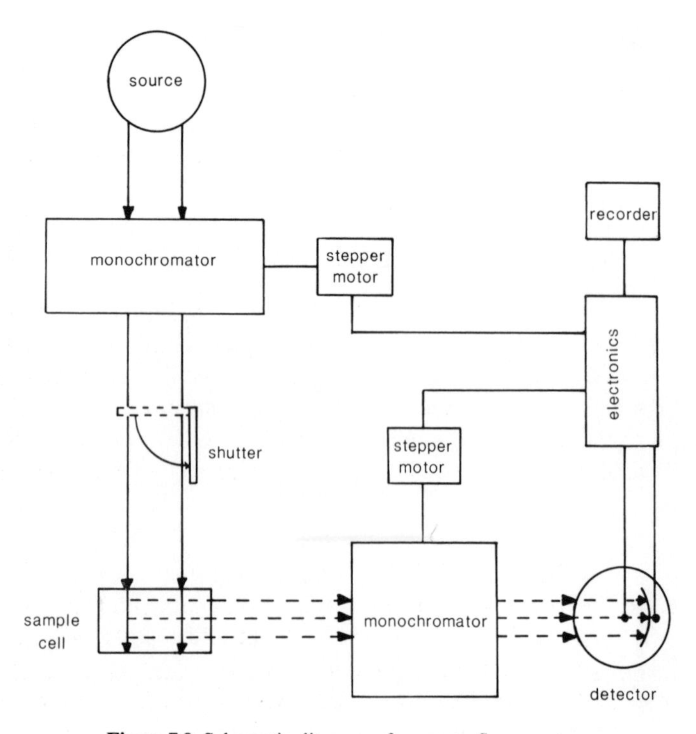

Figure 7.9 Schematic diagram of a spectrofluorometer.

(c) a sample cell
(d) a filter or monochromator to select the emission wavelength
(e) a photodetector
(f) a data-readout device

The radiation is focused by slits and mirrors both in the incident and emitted beams. The key elements of a fluorometer are shown in Figure 7.9. The emitted light is usually measured at 90° to the incident radiation. The presence of a shutter in the incident light beam is advantageous, since the high light intensities used in fluorescence instruments can induce decomposition of samples. Hence the sample should be exposed to the incident radiation for the minimum period of time. All lenses, mirrors and windows are quartz, since glass absorbs UV radiation. Single-beam instruments are more common than double-beam fluorometers, since maximum intensity of the exciting radiation is required for good sensitivity and double-beam studies would involve splitting the incident beam equally into two paths. Corrections for lamp instability are performed in some fluorometers by splitting off a fraction (for example, 5%) of the incident light and allowing it to fall directly onto a reference detector. Photomultipliers are used as detectors in almost all fluorometers. The principles of these detectors have been discussed in Section 6.12.4.

7.6.1 *Light sources*

High-pressure xenon arc lamps are used as the light source in almost all common fluorometers. These lamps provide an intense and relatively stable light output which is continuous in the range 270–700 nm. Xenon atoms are ionized by collision with electrons which flow across an arc under the influence of a direct current. The radiation is emitted when the electrons recombine with xenon ions.

Low-pressure mercury-vapour lamps are often used in filter fluorometers. The arc discharge is much less intense than in a high-pressure arc, but the stability is better. Mercury-vapour lamps emit light at specific wavelengths, rather than as a continuum, and therefore the lamp may be coated with a phosphor to yield an emission which is almost continuous. This is required unless the analyte of interest absorbs light at a mercury-emission wavelength.

7.6.2 *Monochromators*

Most spectrofluorometers incorporate diffraction gratings rather than prisms as monochromators. Emission spectra rarely have line widths less than 5 nm and therefore the resolution of the monochromator is not a primary concern. More important are good transmission efficiency, which is required for the detection of weakly fluorescent samples, and low stray-light levels, which are also necessary for good sensitivity. The dispersion of the monochromator is expressed in nanometres per millimetre where the slit width is in millimetres. The slit width is variable and is selected for each spectrum. Larger slit widths give higher signal-to-noise ratios, but smaller slit widths give improved resolution. Consequently, a compromise slit width must be chosen.

Diffraction gratings may be either planar or concave. Planar gratings are produced mechanically and may contain imperfections in some of the grooves which can lead to stray-light transmission. The best instruments use concave gratings which are produced by holographic and photoresist methods. Monochromators commonly use planar gratings with 600 grooves mm^{-1} with the blaze angle (in the first order) selected for maximum efficiency at 300 nm in the excitation and 500 nm in the emission unit. Filters are used to block out diffracted light of higher order.

7.6.3 *Sample details*

Samples are generally studied as dilute solutions in 10 mm clear-sided cuvettes. Any solvent which does not absorb radiation in regions overlapping the fluorophore's excitation and emission wavelengths can be used. Since the volume of solution observed is usually about 1 μl, small sample cuvettes can be used as long as they are properly aligned in the fluorometer. Fluorometry is a very sensitive technique and solutions of 10^{-8} M are commonly studied. The normal geometry used in fluorescence studies involves the incident radiation passing centrally through one face of the cuvette with measurement of the radiation emitted at 90° through the centre of a second face. However, if the sample is turbid due to suspended particles, or is solid, the cuvette can be rotated to allow the incident beam to impinge on a point near the front of the sample, so that neither the incident nor the emitted radiation is scattered by the turbid sample.

7.6.4 *Filter fluorometers*

Filter fluorometers are relatively inexpensive compared with spectrofluorometers and they may be very convenient for routine quantitative determinations of specific fluorescent molecules. A mercury lamp is usually used as the light source. An absorption or interference filter allows selection of the excitation wavelength band and a second filter located between the sample and the photomultiplier transmits the fluorescent radiation. The use of optical filters gives high sensitivity, since they transmit a relatively wide band of radiation, but the selectivity is correspondingly poor. Filter fluorometers often use a double-beam arrangement to reduce the effects of fluctuations in the light source. If flow cells are incorporated, the instrument can be used as a detector for HPLC.

7.7 Applications of fluorescence spectrophotometry

Although the fluorescence spectra of a molecule may be useful in structure confirmation, fluorescence spectrophotometry plays a minor role in qualitative analysis. However, the sensitivity and specificity of the technique makes it ideally suited to the detection and quantitative determination of a wide range of analytes present at low levels. Many drugs, some vitamins including A, E and riboflavin, and toxic components of food including mycotoxins, can be determined by fluorescence spectrophotometry. In some cases, the analyte must be separated from other components before the determination. The use of a fluorescence detector with HPLC separation is a common and powerful combination. Thus, aflatoxins can be determined at subnanogram levels by this procedure. The fluorometric response for aflatoxin G_2 is 62·5 times that obtained by UV absorbance (Panalaks and Scott 1977). Fluorescent dyes such as 2′, 7′-dichlorofluorescein can be used to detect spots on a TLC plate, either by incorporation into the adsorbent or by spraying a solution onto the developed plate prior to viewing the plate under a UV lamp.

The use of fluorescent labels has facilitated the development of fluorescence immunoassay which is a rapidly expanding field for the analysis of biological molecules in clinical medicine and other areas. An antigen labelled with a fluorescent molecule may exhibit a change in fluorescence properties on binding to an antibody. If the change in the fluorescence on binding is sufficient, the antigen may be quantified without a separation of the bound and unbound labelled molecules. This is known as homogeneous immunoassay. If the bound antigen must be separated from the unbound molecules, the technique is called heterogeneous immunoassay. Various types of fluorescence immunoassay have been developed. Homogeneous techniques based on fluorescence-excitation transfer, fluorescence polarization, enhancement and quenching of fluorescence, and heterogeneous techniques based on the use of fluorescent labelled antibodies have been developed (O'Donnell and Suffin 1979). Immunoassay techniques are discussed in detail in texts by Morris and Clifford (1985) and Tijssen (1985).

The use of fluorescent probes can extend the range of application of fluorescence spectrophotometry to studies on the structural and kinetic properties of macromolecules. A probe is a fluorescent molecule containing a reactive functional group which can be chemically bound to a macromolecule. Probes can be designed to possess emission properties sensitive to particular changes in the local environment, thereby providing structural information about a part of the macromolecule which cannot otherwise be studied by fluorescence. Various fluorescent probes can be chemically bound to specific functional groups in a series of studies and this allows investigations of the polarity and accessibility of the environment of fluorophores in several regions of the macromolecule. Rates of molecular rotation can also be studied with fluorescent probes. A probe perturbs the molecule, however, and checks need to be included to ensure that the perturbation does not negate the validity of the results of any study involving a fluorescent probe. An example of the application of fluorescent probes is given in Section 7.5.2.

References

Brown, S.B. (1980) *An Introduction to Spectroscopy for Biochemists*, Academic Press, London.
Lakowicz, J.R. and Weber, G. (1973) *Biochemistry* **12**, 4171.
Morris, B.A. and Clifford, M.N. (1985) *Immunoassays in Food Analysis*, Elsevier, London.
O'Donnell, C.M. and Suffin, S.C. (1979) *Analyt. Chem.* **51** (1), 33A.
Panalaks, T. and Scott, P.M. (1977) *J. Assoc. Off. Analyt. Chem.* **60** (3), 583.
Shimizu, O. and Imakuvo, K. (1977) *Photochem. and Photobiol.* **26**, 541.
Tijssen, P. (1985) Practice and theory of enzyme immunoassays. Vol. 15 in *Laboratory Techniques in Biochemistry and Molecular Biology*, Burdon, R.H. and van Knippenberg (eds.), Elsevier, Amsterdam.
Vladimirov, Y.A. and Burstein, E.A. (1960) *Biophysika* **5**, 385.
Waggoner, A.S. and Stryer, L. (1970) *Proc. Nat. Acad. Sci. USA* **67**, 579.
Weinryb, I. and Steiner, R.F. (eds.) (1971) *Excited States of Protein and Nucleic Acids*, Plenum, New York.

Further reading

Dyke, K.V. (1985) *Bioluminescence and Chemiluminescence: Instruments and Applications*, Vol. I and II, CRC Press, New York.
Lakowicz, J.R. (1983) *Principles of Fluorescence Spectroscopy*, Plenum, New York.
Schulman, S.G. (1977) *Fluorescence and Phosphorescence Spectroscopy: Physicochemical Principles and Practice*, Pergamon, New York.
Wehry, E.L. (1976–81) *Modern Fluorescence Spectroscopy*, vols. 1–4, Plenum, New York.

8 Infrared spectroscopy

8.1 Introduction

Infrared (IR) spectroscopy covers the region of the electromagnetic spectrum between the red end of the visible range and the microwave region. Standard spectrometers scan the mid-IR region between 4000 and 400 cm^{-1} (2·5–25 μm). The near-IR extends from 12 500–4000 cm^{-1} (0·8–2·5 μm), and the region from 400–50 cm^{-1} (25–200 μm) is known as the far-IR. IR radiation causes transitions between vibrational energy levels and also between rotational energy levels. Transitions between rotational energy levels are manifested in the gas phase, but spectra in the liquid or solid state show broad bands in which the peaks arising from the individual rotational transitions are not resolved. The frequency of absorption due to transitions between vibrational energy levels allows the identification of functional groups within the molecule. The vibrations of molecules may also be usefully investigated by Raman scattering. (Details of this technique are given in 'Further reading'.)

8.2 Molecular vibrations

The atoms of a molecule undergo stretching and bending vibrations with frequencies corresponding to quantized energy levels. Examples of modes of

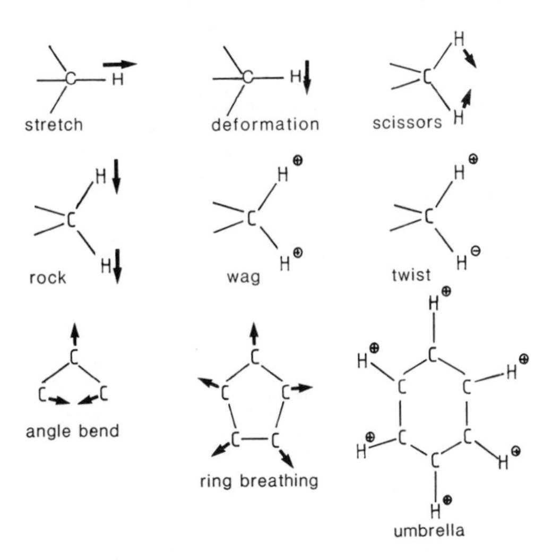

Figure 8.1 Movements in some simple vibrations: + and – refer to motions perpendicular to the plane of the paper.

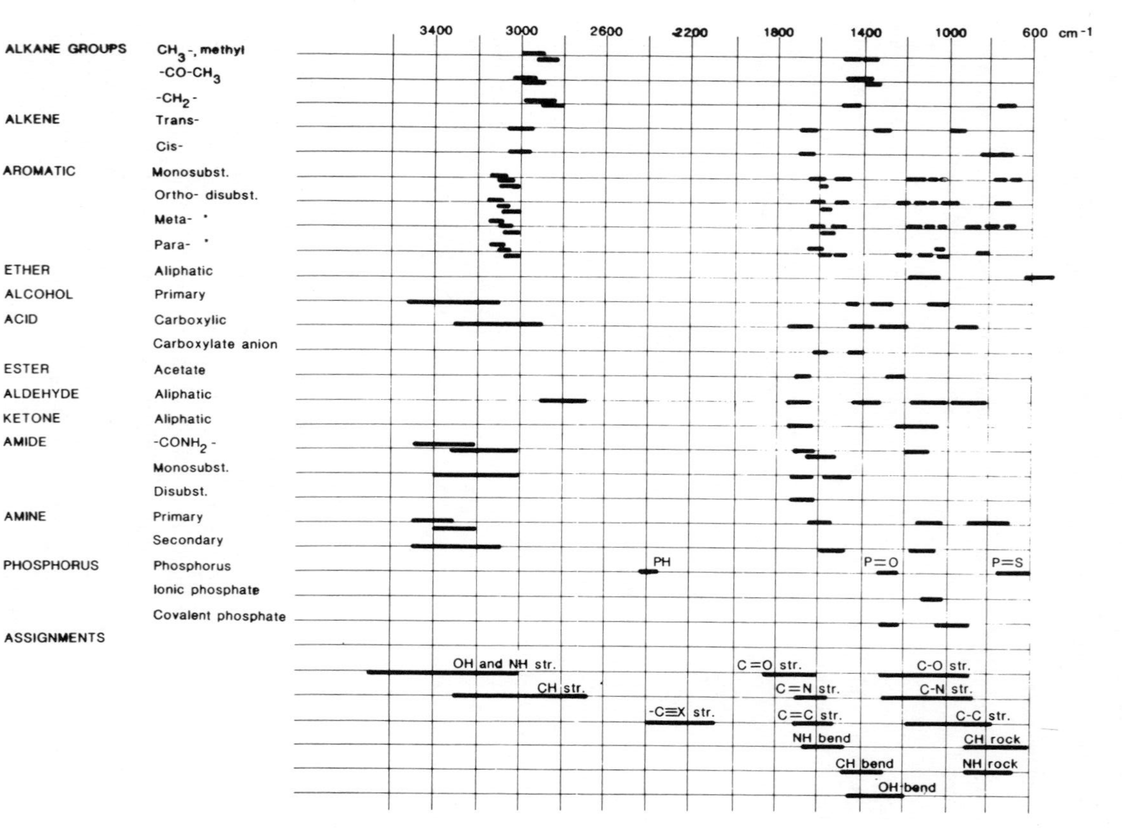

Figure 8.2 Characteristic infrared group frequencies (str. = stretch).

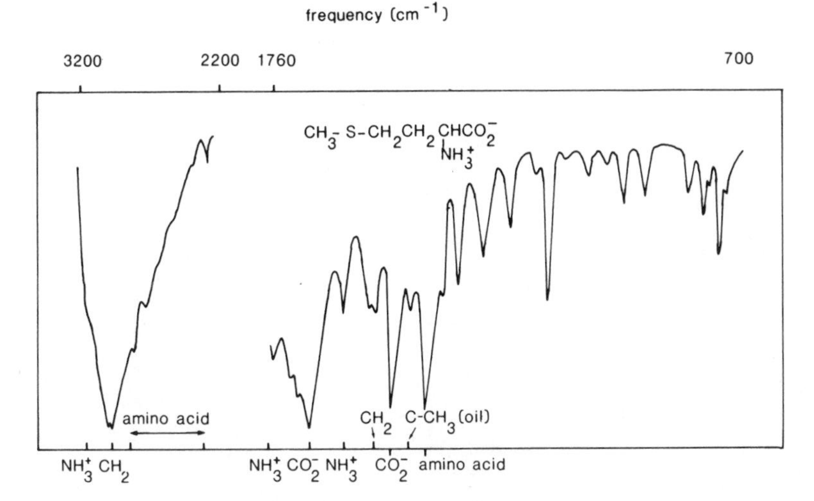

Figure 8.3 Infrared spectrum of *dl*-methionine.

vibration are given in Figure 8.1. When IR radiation at one of these fundamental frequencies impinges on the molecule, energy is absorbed and the amplitude of the appropriate vibration is increased. When the molecule reverts from the excited state to the original ground state, the absorbed energy is released as heat. In order for the radiation to be absorbed, the vibration must involve a change in the dipole moment of the molecule. Although the vibrations are molecular vibrations, the distortion is generally confined to a particular chemical grouping. The frequencies associated with chemical groupings are called characteristic group frequencies and typical values are shown in Figure 8.2. As well as absorption at the fundamental frequencies, absorption bands due to overtones may occur with reduced intensity at multiples of the wave number. Combination and difference bands which are the sum and difference of two or more wave numbers corresponding to fundamental frequencies may also be present, again at reduced intensities. Hence, the IR spectra of biological molecules may become very complex. The spectrum of the amino acid methionine is a typical example (Figure 8.3). Although some of the peaks can be assigned to specific functional groups, many IR absorption bands cannot be interpreted with confidence. Absorption bands due to stretching vibrations can often be assigned, but this is not possible in many cases for the multitude of bands arising from the bending vibrations of an organic molecule.

8.3 Qualitative analysis

A spectrum in the mid-IR region generally allows the identification of some of the functional groups present in a molecule, but full structural determination

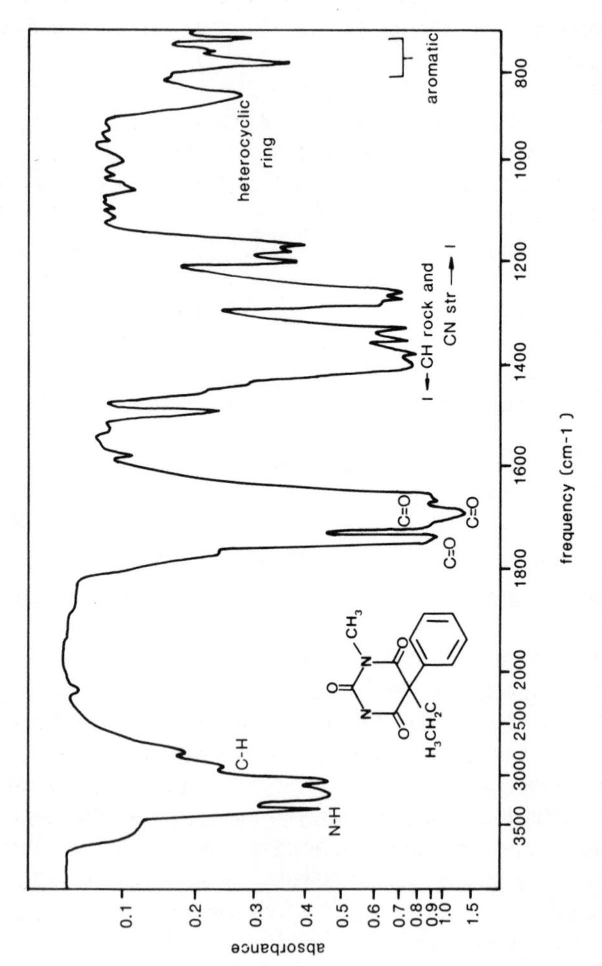

Figure 8.4 IR spectrum of phenobarbitone (KBr disc) Redrawn with permission from Levi and Hubley (1956).

often requires a combination of IR with other techniques. An IR spectrum is particularly useful for the identification of certain chemical groups including alkene, amino, aromatic, carbonyl, carboxy and hydroxy groups. The region between 1500 and 600 cm^{-1} (6·7–16·7 μm) is rich in absorption bands and shoulders due to bending vibrations as well as C—C, C—O, and C—N stretching vibrations. This region is often called the fingerprint region and can be helpful when comparing a spectrum of an unknown with that of a known compound, because molecules of similar structure often have discernible differences in this region.

Assignment of IR absorption bands can be illustrated with the spectrum of the barbiturate phenobarbitone (Figure 8.4). A free N—H vibration occurs at 3335 cm^{-1}, with broad hydrogen-bonded N—H absorption bands at 3200 and 3100 cm^{-1}. The absorptions at 2937 and 2846 cm^{-1} are C—H stretching vibrations, and three bands due to the C=O groups occur at 1772, 1737 and 1710 cm^{-1}. The low-frequency band is associated with the C=O group in the 2 position, which is flanked by two nitrogen atoms. These atoms contribute single-bond character and hence cause a shift to lower frequency due to their stabilization of valence bond isomers II and III shown below.

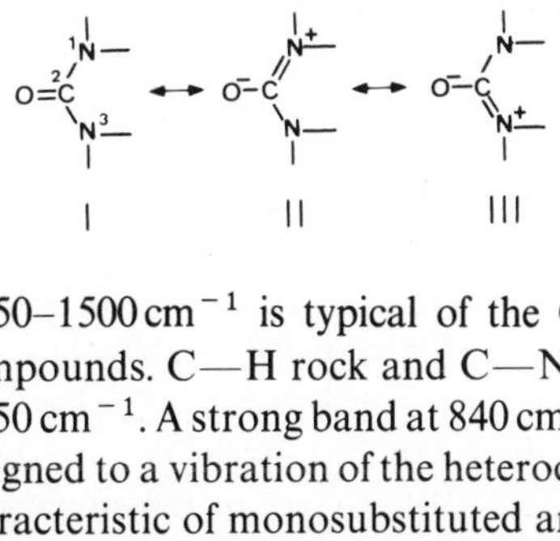

I ↔ II ↔ III

The absorption at 1450–1500 cm^{-1} is typical of the C—C stretching vibrations of aromatic compounds. C—H rock and C—N stretching vibrations are evident at 1470–1250 cm^{-1}. A strong band at 840 cm^{-1} is common to most barbiturates and is assigned to a vibration of the heterocyclic ring. Two strong absorption bands, characteristic of monosubstituted aromatics, occur at 770 and 715 cm^{-1} due to an out-of-plane C—H deformation.

The near-IR region includes many absorption bands arising from overtones or combinations of fundamental frequencies. These include the first overtone of the O—H stretching vibration at 7140 cm^{-1} and the first overtone of the N—H stretching frequency at 6667 cm^{-1}. This region is valuable for quantitative analysis, particularly by reflectance spectroscopy (see Section 8.8).

The far-IR region contains absorptions arising from bending vibrations and is sometimes valuable for studies of organometallic compounds.

8.4 Quantitative analysis

Absorption of IR radiation by molecules is governed by the Beer–Lambert law. Molar absorption coefficients in the range 50–1000 are considered strong, while weak bands have values of about 1. This contrasts with UV and visible spectrophotometry where molar absorption coefficients up to about 30 000 are common. Quantitative analysis by mid-IR spectroscopy is less common than by other spectroscopic techniques, since the accuracy is often limited due to overlap of absorption bands, difficulties of specifying the baseline and errors due to scattered radiation. The determination of *trans*-unsaturated fatty acids in edible fats by their absorption in the IR at 970 cm^{-1} (10.3 μm) is a standard procedure (IUPAC 1975). Another application of IR spectroscopy in quantitative analysis is the determination of ethanol in breath by the absorption at 2950 cm^{-1} (3.4 μm). The Lion Intoximeter, used by the police to detect alcohol in the breath of motorists in the United Kingdom, operates on this principle. IR spectroscopy is also used in the food industry for the quantification of fat, protein and lactose in milk using the absorptions at 1745 cm^{-1} (5.7 μm), 1550 cm^{-1} (6.5 μm) and 1040 cm^{-1} (9.6 μm) respectively.

8.5 Instrumentation

A mid-IR transmission spectrometer comprises a radiation source, monochromator, sample cell and detector. The optical components cannot be made of glass, and salts are commonly used. The standard spectrometer for studies in the mid-IR spectral region is a double-beam instrument in which radiation is passed alternately through the sample and reference cells.

8.5.1 *Sources*

Sources that radiate a continuous spectrum approximating to a black body are generally used. A Nernst Glower, which is a hollow rod of mixed rare-earth oxides electrically heated to 1500–1900 °C, represents an excellent source for wavelengths in the range 4000–1000 cm^{-1} (2·5–10 μm), but the radiation is less intense below 1000 cm^{-1}. The Globar, comprising a rod of silicon carbide operating at 1200–1300 °C, is a less intense source than the Nernst Glower over most of the mid-IR region, but is more intense below 650 cm^{-1} (15·4 μm). An alternative source which is simple, rugged and reliable is a coil of Nichrome wire. This is operated at temperatures up to 1100 °C, but it is less intense than the other sources mentioned above.

8.5.2 *Monochromators*

Most modern IR spectrometers employ a diffraction grating rather than a prism for the production of a narrow wavelength band. The principles are similar to those of monochromators used in UV and visible spectrophotometers as described in Section 6.12.2. Gratings for use in the IR region may be up to 20 cm square with 300–3000 lines cm^{-1}. The resolution of a typical IR spectrometer is about 2 cm^{-1}.

8.5.3 *Detectors*

The detectors used in the mid-IR region fall into two categories: thermal detectors and quantum detectors. Several types of thermal detectors are used, including thermocouples, thermopiles, bolometers or pyroelectric detectors. Thermal detectors rely on IR radiation falling on a small piece of blackened gold or similar material and causing an increase in temperature which is detected and converted into an electrical signal by the sensor. Temperature changes of a few thousandths of a degree must be detectable. A highly sensitive thermocouple or a thermopile, which consists of several thermocouples in series, can be used as the sensor. These consist of junctions of dissimilar metals, such as bismuth and antimony, mounted in a vacuum. The thermocouple generates a potential of a fraction of a microvolt in response to the small temperature rise.

Bolometer detectors consist of a metal filament or semiconductor crystal whose electrical resistance varies with temperature. The Golay detector relies on the expansion of a gas in a blackened receiver chamber heated by the IR radiation. One wall of the chamber consists of a flexible diaphragm which is silvered on the outside and as the gas expands, the diaphragm distorts. Light from a source is reflected from the silvered surface of the diaphragm and the distortion reduces the intensity of the light reflected onto a photocell. This device is very sensitive and reliable. The pyroelectric detector is a crystal such as barium titanate which develops a potential across its faces when it is heated. This detector has a good response to radiation across a wide range of IR frequencies and intensity.

Quantum detectors are made of semiconducting materials which have an electronic structure such that IR photons may excite an electron from a fully occupied molecular orbital to an empty orbital in which it becomes conducting. A current proportional to the IR radiation will flow on application of a potential across the crystal. Semiconducting materials include indium antimonide, lead sulphide and germanium. Quantum detectors are very sensitive, but they can only be used over limited wavelength ranges. The sensitivity and range of the detectors may be improved by cooling them in liquid nitrogen.

8.5.4 *Double-beam spectrometers*

The use of a double-beam instrument has several advantages. The energy of the radiation source is frequency-dependent and the signal level is low and subject to noise. Chopping the radiation between the sample and reference allows accurate determination of the ratio of the intensity of the radiation transmitted by the sample to that transmitted by the reference. Interference from carbon dioxide and water vapour is reduced and a relatively stable alternating current amplifier can be used. The noise level can be reduced by

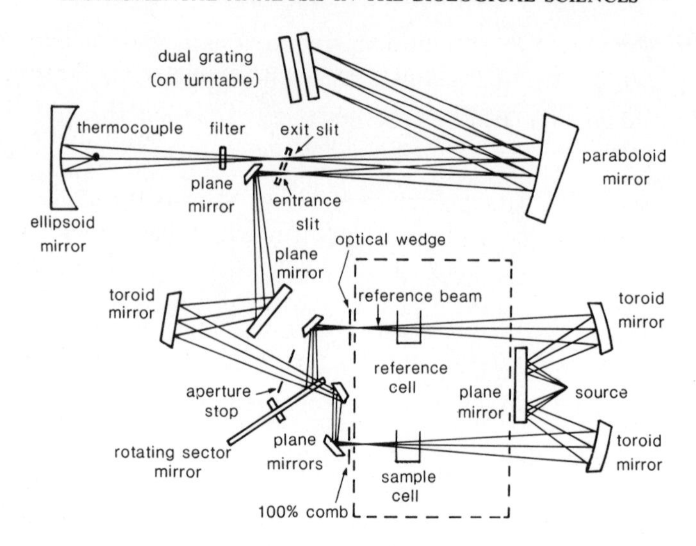

Figure 8.5 Optical diagram of a filter-grating double-beam IR spectrophotometer.

tuning the amplifier to the chopping frequency with rejection of all other frequencies.

A schematic diagram of a double-beam instrument is shown in Figure 8.5. The source emits radiation which is split into a sample beam and a reference beam. An attenuator, comprising a comb-shaped shutter or optical wedge, is moved in and out of the reference light path by a motor controlled by the amplifier detector signal. The motor drives the attenuator so as to equalize the intensity of radiation reaching the detector through the sample and reference paths. A rotating sector mirror or chopper is used to send the sample beam and reference beam alternately through the monochromator to the detector. The percentage transmission is proportional to the distance moved by the attenuator and is recorded on a chart. Scanning of the spectrum is achieved by a motor which rotates the prism or grating monochromator and also drives the chart. This mode of operation is described as an optical-null system. The major drawback is that intense absorption by the sample will cause the attenuator to block off practically all the light in the reference beam. This can cause the recorder pen to drift misleadingly.

8.5.5 *Fourier transform infrared spectroscopy* (*FT–IR*)

An interferometer combined with Fourier transform data analysis represents an alternative to a diffraction grating for resolving IR radiation into its component wavelengths. The use of an interferometer is advantageous in the far-IR region, where the source is less intense and diffraction gratings cover a narrow range of the spectrum. Additionally, there are advantages in the mid-

IR region where rapid data acquisition is required, as in an IR detector for gas chromatography. There are several types of interferometer, but the simplest is the Michelson interferometer (Figure 8.6). Radiation from a source is collimated by mirror A and the beam is reflected on to a beam splitter, which reflects half the radiation, beam 1, on to mirror B, and allows the rest of the radiation, beam 2, to pass through to mirror C. Mirrors B and C reflect the radiation back to the beam splitter, which again reflects half the radiation, and transmits half. Thus, half of each light beam passes through the sample to the detector. For light of wavelength λ, beams 1 and 2 will be in phase at the detector if the path lengths are equal or differ by a multiple of λ. Constructive interference occurs if this condition is met, while destructive interference occurs at half-integer multiples of λ.

Mirror C is moved towards the beam splitter during the analysis. Since the source emits radiation over a range of wavelengths, the detector signal varies in a complex manner as mirror C is moved. Absorption of radiation at specific wavelengths by the sample modifies the variation of signal with wavelength. The resultant pattern is called an interferogram, and can be converted by Fourier transformation into a spectrum. Fourier transformation is a mathematical technique involving the separation of the interferogram into a combination of a large number of sine and cosine terms varying in frequency. Energy-wasting slits are avoided by the use of an interferometer and the signal-

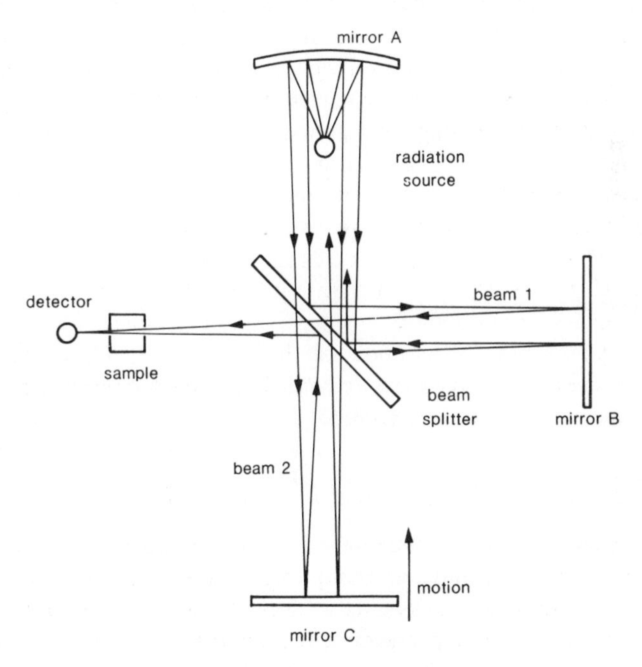

Figure 8.6 The Michelson interferometer.

to-noise ratio and resolution are very good. The spectrum is obtained rapidly compared with a traditional IR spectrometer.

8.6 Sample presentation

Glass absorbs strongly in the IR region and therefore cells are generally made of sodium chloride (transparent above 670 cm^{-1}) or other salts including potassium bromide (transparent above 380 cm^{-1}). Liquids may be studied undiluted as a thin film between salt discs or as solutions in a suitable solvent. Carbon tetrachloride, carbon disulphide or chloroform are commonly used as solvents, since they have few absorption bands in the IR region. IR spectrometers are usually double-beam instruments, with a solvent blank in the reference beam. This procedure minimizes the parts of the spectrum that are obscured by solvent absorption and also eliminates the effects of absorption by carbon dioxide and water vapour in the light path. Cells for solution spectra commonly have a path length of 0·1 mm, with solutions of 10% concentration typically being used.

Since water cannot be used as a solvent, biological materials in the solid state are best determined as an alkali halide pellet. About 1 mg of a dry sample and 100 mg of a salt such as potassium bromide are ground together and pressed into a small disc under high pressure. The resulting pellet comprises a solid solution of the material in the salt. The spectra of solids may also be determined as a thin film of a mull or paste prepared by mixing the material with a few drops of the hydrocarbon, Nujol. Nujol mulls give no information in regions of the spectrum where the hydrocarbon itself absorbs strongly.

The IR spectra of gases may be determined by using cells with long path lengths, from 10 cm up to several metres. The use of internal mirrors can extend the effective path length of a cell without making the dimensions of the cell unmanageable. Sensitivity is also improved by studying gases under increased pressure.

8.7 Attenuated total reflectance

Complex solid samples can be studied in the IR by the technique of attenuated total reflectance (ATR), also known as multiple internal reflectance. When the angle of incidence of radiation at the surface of a prism is small (Figure 8.7*a*), partial refraction and partial reflection occurs. However, when the angle of incidence exceeds a critical value, total internal reflectance occurs at the surface of the prism (Figure 8.7*b*). In practice, the radiation penetrates a short distance into the less dense medium and this penetrating radiation can be absorbed by placing a sample in intimate contact with the prism (Figure 8.8). The reflected radiation then yields an absorption spectrum which closely resembles a transmission spectrum of the sample.

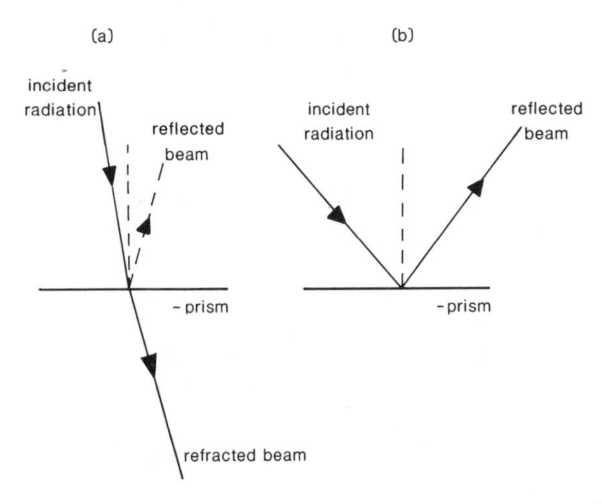

Figure 8.7 (*a*) Partial reflection and partial refraction at low angles of incidence; (*b*) total internal reflection at higher angle of incidence.

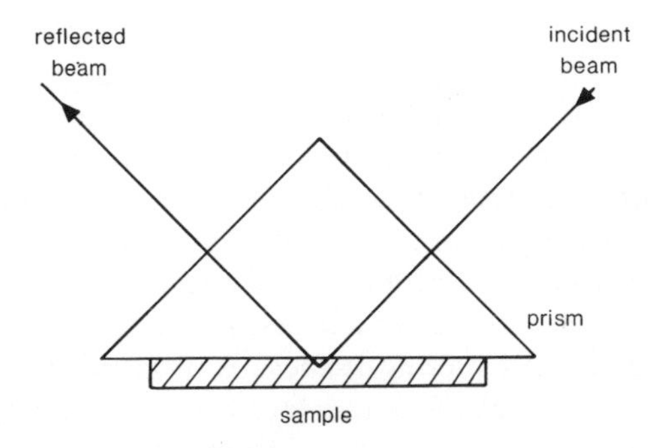

Figure 8.8 Schematic diagram of attenuated total reflectance in a single reflection prism.

8.8 Near-infrared reflectance analysis

Near-IR reflectance analysis has developed rapidly in recent years and has become an important technique for the quantitative measurement of a wide variety of components in solid materials. Protein, moisture, oil, fibre and starch in food and agricultural materials, moisture and active ingredients in pharmaceuticals, and moisture and nicotine in tobacco are among the applications of the technique.

The near-IR spectral region is dominated by weak overtones and combinations of vibrational bands from the mid-IR region. All stretching bands involving hydrogen atoms are represented and the high density of absorption frequencies in the region leads to several broad peaks, as can be seen in the

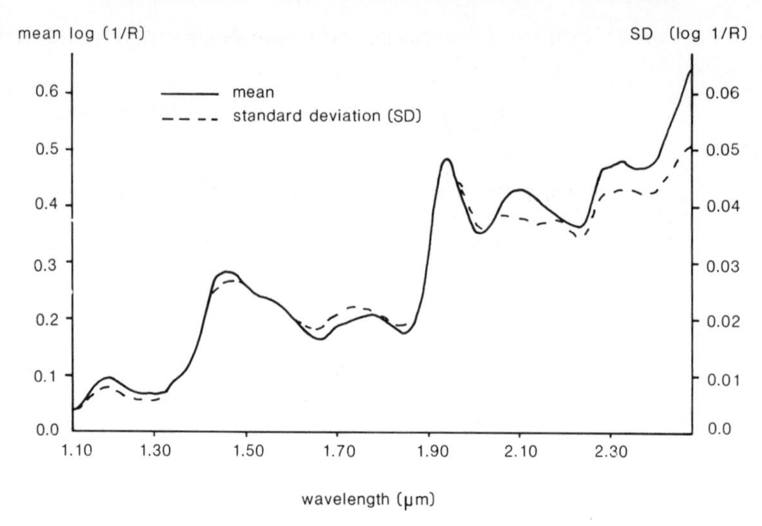

Figure 8.9 Mean and standard deviation near-infrared spectrum for a sample of wheat flour. Redrawn with permission from Cowe and McNicol (1985).

spectrum of wheat flour (Figure 8.9). Although absorption intensities in the near-IR are orders of magnitude weaker than those in the mid-IR, the tungsten-filament lamp is an intense source of radiation, and highly sensitive detectors such as lead sulphide are considerably more sensitive than those in the mid-IR. Many of the sources of error in mid-IR quantitative analysis are associated with high sample absorption and these can be avoided in the near-IR. In addition, the light-scattering coefficients are larger at the relatively short wavelengths of the near-IR region and, together with lower absorption levels, this leads to a high ratio of scattered to absorbed radiation. The reflectance response is linearly proportional to component concentration at high levels of scattering. Thus, the near-IR region has several important features for the analysis of components in complex solid materials. Analysis requires interpretation of the data by correlation transform spectroscopy, which involves multiplying the absorbance at several selected wavelengths by a specific coefficient for each of the components being measured. The procedure is an empirical method which uses computer regression analysis to select the wavelengths and the correct coefficients applicable to a particular sample matrix. The spectra of the components are considered when selecting appropriate wavelengths for the calculation.

Near-IR reflectance analysis requires little sample pretreatment; the sample is usually presented as a liquid or a solid ground into a relatively coarse powdered form. The technique is rapid after the initial calibration and is non-destructive. Although near-IR is a secondary procedure which is dependent on calibration against other standard methods, the advantages of the technique make it very suitable for routine analysis and it is applied widely in industrial laboratories.

References

Cowe, I.A. and McNicol, J.W. (1985) *Appl. Spectroscopy* **39**(2), 257.
Levi, L. and Hubley, C.E. (1956) *Analyt. Chem.* **28**(10), 1591.
IUPAC (1979) *Standard Methods for the Analysis of Oils, Fats and Derivatives*, 6th edn, Pergamon, Oxford.

Further reading

Brame, E.G.Jr. and Grasselli, J.G. (1976–77) Infrared and Raman spectroscopy, in *Practical Spectroscopy Series*, vol. 1, parts A, B and C, Marcel Dekker, New York.
Colthup, N.B., Daly, L.H. and Wiberley, S.E. (1975) *Introduction to Infrared and Raman Spectroscopy*, 2nd edn, Academic Press, New York.
Nakanishi, K. and Solomon, P.H. (1977) *Infrared Absorption Spectroscopy*, 2nd ed, Holden-Day, San Francisco.
Smith, A.L. (1979) *Applied Infrared Spectroscopy*, Wiley, New York.
Theophanides, T.M. (1979) *Infrared and Raman Spectroscopy of Biological Molecules*, Reidel, Dordrecht.

9 Nuclear magnetic resonance spectroscopy

9.1 Introduction

Nuclear magnetic resonance spectroscopy (NMR) is a technique that has developed rapidly from early experiments in the late 1940s until the present day, when it plays a key role in structural analysis and identification of organic molecules. Although NMR is less sensitive than some techniques, such as UV or fluorescence spectrophotometry, it can be applied much more widely and provides specific structural information that is often adequate for the unambiguous identification of organic molecules.

NMR depends on the absorption of radiation by the nuclei of certain isotopes. The technique requires a strong magnetic field to remove the degeneracy of the nuclear energy levels and a radiofrequency (rf) field to induce transitions between the energy levels. Isotopes with suitable nuclear magnetic properties include ^{1}H, ^{13}C, ^{15}N and ^{31}P.

9.2 Principles

9.2.1 *Nuclear energy levels*

Many atoms have an inherent angular momentum and magnetic moment which can be interpreted in a classical model as arising from the spinning of the nucleus about its axis. When the spinning nucleus is placed in an external magnetic field B_0, the nuclear magnetic moment precesses about B_0 (Figure 9.1) in the same way as a spinning top precesses about the earth's gravitational field. For most nuclei, there are two or more quantized energy levels for the nucleus in a magnetic field. These energy levels arise from the interaction between B_0 and the magnetic moment μ of the nucleus, which is tilted at an angle to the applied magnetic field.

$$\mu = \gamma \frac{h}{2\pi}[I(I+1)]^{\frac{1}{2}}$$

where γ is the magnetogyric ratio, h is the Planck constant and I is the nuclear spin quantum number. Each allowed value of I corresponds to a discrete orientation (and energy level).

NMR is applied most commonly to nuclei for which $I = \frac{1}{2}$. These nuclei include ^{1}H, ^{13}C, ^{15}N and ^{31}P, and for these isotopes there are two quantized

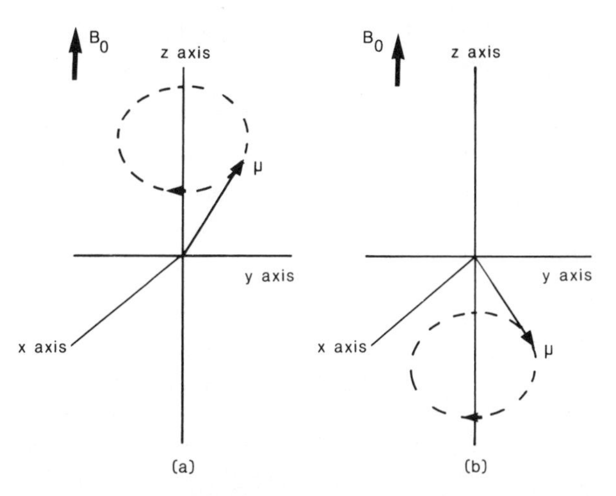

Figure 9.1 Precession of nuclear magnetic moment, μ, about the external magnetic field, B_0. (*a*) Lower-energy state; (*b*) higher-energy state.

energy levels which can be visualized as corresponding to alignment of the magnetic moment, either with or against the magnetic field (Figure 9.1). The discussion in this chapter will be restricted to these isotopes. Other isotopes have $I = 0$, or $\geqslant 1$. Isotopes with $I = 0$, including ^{12}C, have zero nuclear magnetic moment and hence do not produce an NMR spectrum, while isotopes with $I \geqslant 1$ have $2I + 1$ nuclear energy levels in a magnetic field. Such isotopes include ^{2}H, ^{14}N, ^{35}Cl and ^{39}K, and these nuclei have a non-spherical nuclear-charge distribution and hence an electric quadrupole moment which leads to complexities in the NMR spectrum.

The rate of precession of the nuclear moment corresponds to the Larmor frequency ν_0, where

$$\nu_0 = \frac{\gamma}{2\pi} B_0$$

The energy E of the nuclear magnetic energy levels is given by

$$E = \gamma \frac{h}{2\pi} m B_0$$

where m is the magnetic quantum number and may take the value $-\frac{1}{2}$ or $+\frac{1}{2}$ if $I = \frac{1}{2}$.

At equilibrium, nuclei will be distributed between the two energy levels according to the Boltzmann distribution

$$\frac{n_{\text{upper}}}{n_{\text{lower}}} = \exp(-2\mu B_0/kT)$$

Consequently, when a magnetic field is applied to the sample, there will be more nuclei aligned with B_0 than opposed to it and there is a net macroscopic magnetization in the direction of the field.

9.2.2 *Magnetic resonance*

If a radiofrequency (rf) magnetic field B_1 is applied in a plane at 90° to B_0, it induces transitions between the two energy levels if the frequency corresponds to the Larmor frequency of the nucleus. Because the number of nuclei in the lower energy level exceeds that in the upper energy level, there is a net absorption of energy at this resonance frequency ν_1, which is recorded as the NMR signal.

The resonance frequency varies widely for different nuclei. At an applied field strength of 2·349 T, the values of ν_1 correspond to 100 MHz for ^{1}H, 25·145 MHz for ^{13}C, 40·481 MHz for ^{31}P and 10·137 MHz for ^{15}N. Although the resonance frequency of an atom is dependent on its chemical environment, all the atoms of a particular isotope have resonance frequencies within a relatively narrow range and hence an NMR spectrum arising from the nuclei of a single isotope can be recorded.

9.2.3 *Relaxation processes*

After the nuclei of an isotope have absorbed energy at the resonance frequency, no further signal is observed if the populations of the upper and lower energy levels become equal. This is known as saturation of the signal. However, there are two relaxation processes which cause nuclei in the upper energy level to relax back to the lower energy level in a non-radiative transition.

The first relaxation process, spin–lattice or longitudinal relaxation, is a first-order process characterized by the lifetime T_1. Spin–lattice relaxation relates to relaxation of the magnetization parallel to B_0. It arises from the interaction of the nuclear spin with the fluctuating magnetic fields produced by random motions of neighbouring nuclei. The value of T_1 is usually in the range 0·01–100 s for liquids of low viscosity, but is much longer for solids and viscous liquids. These differences reflect the importance of Brownian motion in the spin–lattice relaxation of nuclei in the liquid state.

The second relaxation process is called spin–spin or transverse relaxation, which refers to the loss of magnetization in the plane perpendicular to B_0. The rf field B_1 causes the nuclear spins to precess in phase after absorption of radiation. However, the phase coherence of the spins is lost through dipole–dipole interactions with neighbouring nuclei. This relaxation process is also first order and is characterized by the lifetime T_2.

The minimum width of an NMR line at half-height, $\nu_{1/2}$, is dependent on the time a nucleus remains in a given energy level, since according to the Heisenberg uncertainty principle

$$\Delta E \Delta t \geqslant h$$

Since relaxation by dipole–dipole interactions is at least as fast as by spin–lattice relaxation, T_2 can be defined in terms of the width of peaks at half-height:

$$\nu_{1/2} = \frac{1}{\pi T_2}$$

For non-viscous fluids, where diffusion and rotation of molecules occur rapidly, $T_1 \approx T_2$, but for solids and viscous liquids $T_1 \gg T_2$.

9.2.4 *Principles of NMR measurement*

9.2.4.1 *Continuous-wave NMR.* Until the last few years, most NMR spectrometers operated on a continuous wave (CW) principle. Each nucleus absorbs radiation at the resonance frequency ν_1, which is dependent on the strength of B_0. Therefore, either B_0 may be varied for a fixed value of ν, or ν may be varied for a fixed B_0 to record the absorption of atoms in different chemical environments, which corresponds to the NMR spectrum. Instrumentally, it has proved easier to operate at a fixed value of ν and to sweep through a range of values of B_0. The term continuous-wave spectroscopy arises from the continuous application of the rf during observation of the spectrum. The rate of scanning the spectrum must be sufficiently slow to record the narrow absorption lines and maintain equilibrium in the sample and the power must be sufficiently low to avoid saturation of the signal. The plot of radiation absorbed against B_0 or ν produces peaks which are Lorentzian in shape, and the area under a peak is proportional to the number of nuclei absorbing at that field and frequency. CW NMR spectrometers are relatively inexpensive, but the recording of a spectrum is relatively slow, since the resonances are detected sequentially, and the sensitivity is not adequate to determine the spectra of most nuclei other than ^1H. A highly sensitive technique is required for other nuclei, such as ^{13}C, because they occur at low

Table 9.1 Nuclear properties

Nucleus	*Spin*	*Relative signal per nucleus (S)*	*%Natural abundance (A)*	$S \times A$
^1H	1/2	1·00	100	100
^2H	1	$9{\cdot}65 \times 10^{-3}$	0·015	$1{\cdot}4 \times 10^{-4}$
^3H	1/2	1.21	—	—
^{13}C	1/2	$1{\cdot}59 \times 10^{-2}$	1.1	0.017
^{14}N	1	$1{\cdot}01 \times 10^{-3}$	99·6	0.10
^{15}N	1/2	1.04×10^{-3}	0.37	$3{\cdot}8 \times 10^{-4}$
^{17}O	5/2	$2{\cdot}91 \times 10^{-2}$	0.04	$1{\cdot}2 \times 10^{-3}$
^{31}P	1/2	$6{\cdot}63 \times 10^{-2}$	100	6.6

natural abundance and also because they produce a weaker signal per nucleus because of the lower value of the magnetogyric ratio (see Table 9.1).

9.2.4.2 *Pulse NMR*. In order to overcome the limitations of CW measurements, pulse NMR spectrometers have been developed. These involve the application of a short rf pulse which excites all the nuclei of an isotope simultaneously. A pulse width of t s will excite all nuclei over a range of $(2t)^{-1}$ Hz. Thus a 10 μs pulse will excite nuclei within ± 25 kHz from the pulse frequency. The nuclear-spin system is observed after the rf pulse has been turned off. The pulse induces a magnetization arising from the excitation of all nuclei of a particular isotope, and the decay of the magnetization with time can be used to determine T_1 or T_2. Analysis of the induced magnetization by the mathematical technique of Fourier transformation allows the spectrum to be determined.

In order to understand pulse NMR it is helpful to consider the magnetization with respect to a coordinate system that rotates about the z-axis at the Larmor frequency ν_0 as shown in Figure 9.2. In the fixed coordinate system, the application of B_0 causes the nuclear magnetization to precess about the

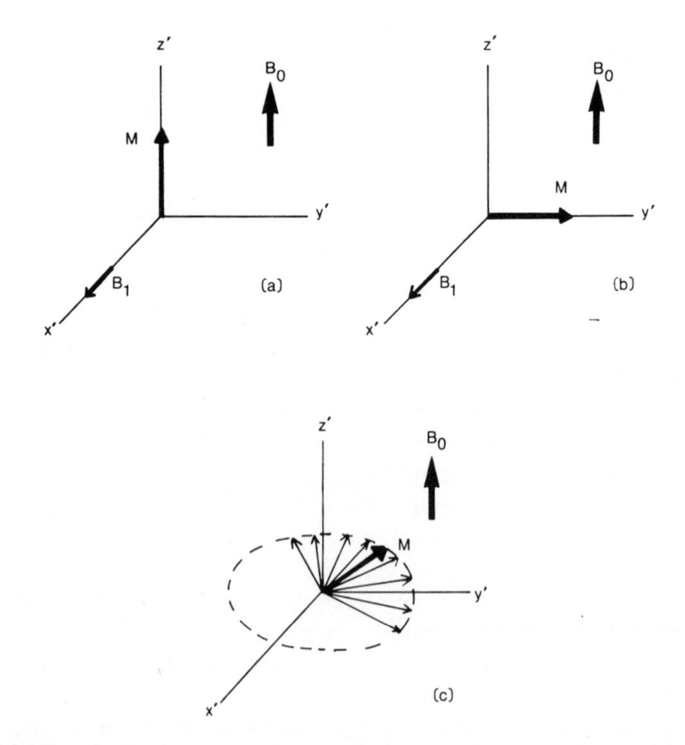

Figure 9.2 Magnetization in a coordinate system rotating at the Larmor frequency along the z axis. (*a*) Before the rf pulse, B_1; (*b*) after a 90° pulse; (*c*) after B_1 has been switched off and some relaxation has occurred both in the $x'y'$ plane due to spin–spin relaxation and towards the z' axis due to spin–lattice interactions.

z-axis with a frequency ν_0 which can be resolved into vector components along the z-axis and in the xy-plane. In the rotating frame, the components of the magnetization in the $x'y'$-plane are zero and the total magnetization only has a component along the z'-axis. If an rf pulse of frequency ν_0 rad s^{-1} is applied in the plane perpendicular to B_0, it can be considered to rotate at this frequency in the fixed coordinate system. In the rotating frame, however, we can arbitrarily apply the rf pulse along the rotating x'-axis. Since it is rotating at the same frequency as the frame, the pulse tips the magnetization from the z'-axis through an angle θ, so that the magnetization has components along the z'- and y'-axes. If the pulse is applied for a time t_p,

$$\theta = \gamma \nu_0 t_p$$

The pulse can be applied for a time sufficient to tip the magnetization from the z'-axis through 90° into the y'-axis. This is called a 90° pulse. After the 90° pulse is turned off, spin–spin relaxation causes the nuclei to exchange energy and the magnetic moments lose their coherence in the $x'y'$-plane. The magnetization along the y'-axis, $M_{y'}$, decays until there is no magnetization in the $x'y'$-plane. Since the magnetic field is not perfectly homogeneous, there are variations in the precession frequency of the nuclei and this also contributes to the decay of $M_{y'}$. The time constant T_2^* for the decay of $M_{y'}$ relates to T_2 and inhomogeneities in the magnetic field, ΔB_0, according to the equation:

$$\frac{1}{T_2^*} = \frac{1}{T_2} + \gamma \frac{\Delta B_0}{2}$$

The nuclear moments also lose energy to their surroundings after the pulse is terminated and this causes the magnetization to relax towards the equilibrium value of M_0 with the time constant T_1. The magnitudes of the time constants are in the order $T_1 \geqslant T_2 \geqslant T_2^*$. Pulse NMR instruments allow detection of the magnetization along the y'-axis. The decay of $M_{y'}$ with time is known as the free induction decay (FID). The FID is the means by which the magnitude and characteristics of M can be monitored. If the rf is precisely at the Larmor frequency, a smooth exponential curve is obtained (Figure 9.3*a*). Since the spectrum of the sample arises from nuclei resonating at various frequencies, interference effects cause a modulation of the simple FID (Figure 9.3*b*).

The FID which can be determined after a single 90° pulse contains information about the absorption of radiation by the nuclei. The decay of the magnetization with time can be converted by Fourier transformation, into an NMR spectrum which indicates the variation of absorption with frequency. Fourier transformation involves the separation of the FID into a combination of sine and cosine terms varying in frequency.

Pulse NMR has advantages over CW NMR for recording spectra, because it is much faster and the signal-to-noise ratio of the process is much better. All nuclei of an isotope are excited simultaneously and thus they produce coherent signals. Multichannel detection in pulse NMR allows an improvement in signal-

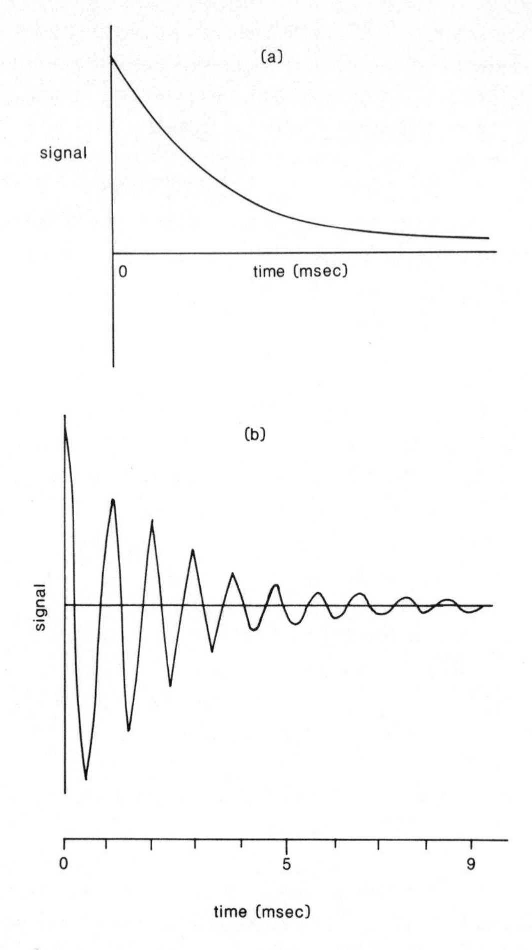

Figure 9.3 Free induction decay after a 90° rf pulse, (*a*) at the Larmor frequency and (*b*) at a slightly different frequency.

to-noise ratio compared with CW NMR up to a limit which corresponds to the total spectral width divided by the typical line width. The improvement is generally in the range 10–20 for protons where the spectrum covers 1 kHz (at 2·3 T), but is greater for nuclei, such as ^{31}P and ^{13}C, where larger spectral widths occur. Summing the FID from a number of pulses also improves the signal-to-noise ratio. Pulse NMR can also be used for determining the relaxation time constants T_1 and T_2. These parameters are measured by the use of various pulse sequences. Determination of T_1 and T_2 is valuable, because these constants provide information about the extent of molecular motion and molecular flexibility. The binding of paramagnetic ions, which reduces the relaxation times dramatically, can also be investigated.

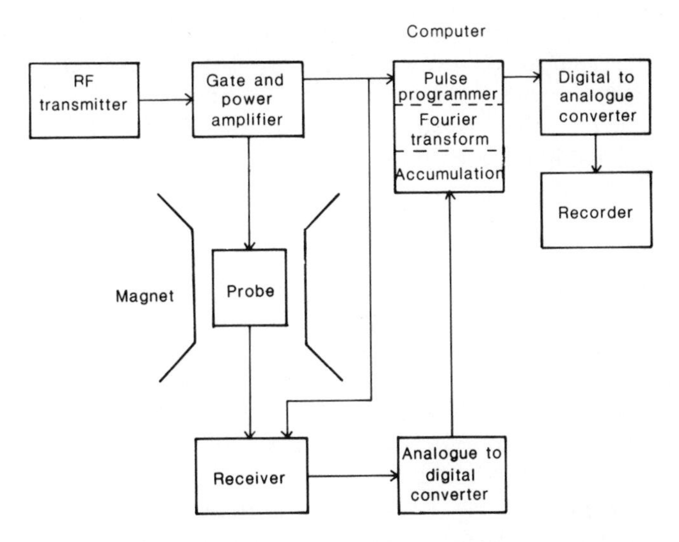

Figure 9.4 Block diagram of a typical pulse NMR spectrometer.

9.3 The pulse NMR spectrometer

A pulse NMR spectrometer (Figure 9.4) consists of six basic elements: the magnet, rf frequency generator, probe, field/frequency lock, computer and recorder.

9.3.1 *The magnet*

The strength of the magnetic field B_0 determines the Larmor frequencies of the nuclei. A high degree of homogeneity and stability is required, since resonance lines are broadened by inhomogeneities. The magnet used may be permanent, electromagnetic or cryogenic. Permanent magnets have high inherent stability if accurately thermostatted, but they are limited to fields up to 2·1 T (corresponding to 90 MHz for ^{1}H). Electromagnets are useful in the 1·8–2·3 T field range (80–100 MHz) and can take larger sample tubes, which improves the signal-to-noise ratio. High currents (of about 50 A) are used, and these must be accurately stabilized. In addition these large currents produce considerable heat, so the coils must be cooled with a rapid flow of thermostatted water. Superconducting magnets have been developed for high fields and large field volumes. A current is established in a coil of niobium–titanium alloy wire with an air core at 4·2 K, the temperature of boiling helium. The coil has zero electrical resistance at this temperature and the current is maintained indefinitely. Superconducting magnets of up to 14 T (600 MHz) are used in NMR spectrometers.

The homogeneity of these magnets is improved by the use of shim coils. These are small coils wound on to formers attached to the pole pieces or the probe. The current in the shim coils can be adjusted to give a specific field gradient which overcomes inhomogeneities in the magnet. The sample tube is also spun at several hundred revolutions per minute during the experiments, to reduce the effects of residual inhomogeneity.

9.3.2 *Rf frequency generator*

A spectrometer designed to measure a variety of nuclei must be capable of generating a wide range of frequencies (100 MHz for ^{1}H, but only 10 MHz for ^{15}N at 2·3 T). It must also produce the frequencies necessary for locking the spectrometer and for any double-resonance experiments that are required. All these frequencies are usually generated by frequency synthesis using a single source to optimize frequency and phase stability. The spectrometer also contains units to amplify these frequencies and gates which control the application of the pulse and detection of the signal.

9.3.3 *Probe unit and sample*

The probe is the sensing unit of the spectrometer. It is located between the pole faces of the magnet. The probe unit houses the sample, the rf transmitters, the output attenuator, the receiver and the phase sensitive detector. The sample usually consists of about 1 ml of liquid in a cylindrical glass tube of about 3 mm i.d. Only a length of 5 mm is in the detector coil, but a column of liquid about 2 cm long is required to avoid end effects. ^{1}H NMR spectra are typically run on samples of about 0·2% in a deuteriated solvent such as $CDCl_3$.

The probe may be a single-coil or crossed-coil type. In crossed-coil probes, one coil supplies rf radiation to the sample and a second coil is mounted at right angles to the first one. A current is induced in the second coil during the experiment and this can be detected and analysed. The single-coil probe supplies and detects rf radiation in one coil. An rf bridge is used to separate NMR absorption from the imposed rf field.

9.3.4 *Field/frequency lock*

In order to minimize the effects of magnetic field fluctuations, the field is locked to the rf frequency by means of servo loops. The NMR signal is an alternating current signal consisting of two components 90° out of phase. The use of a phase-sensitive detector allows only one component to be used for detection, while the second component is used for field-frequency control. The lock used may be either internal or external. An internal lock uses the sample under analysis (often deuterium nuclei in the solvent) for the locking signal, and an external lock uses a signal from a second adjacent sample, such as water or tetramethylsilane in a glass capillary. The reference nucleus is continuously irradiated at its resonance frequency during the experiment. If the field changes, the resonance frequency of the reference nucleus changes and a signal can be input into the shim coils of the magnet until the field is restored and resonance is regained.

9.3.5 *Computer*

The signal from the detector is digitized by an analogue to digital converter and stored in the computer. The computer can average the data from successive FIDs and perform the Fourier transformation required to yield the spectrum. Other mathematical manipulations can also be performed to improve the signal-to-noise ratio. The spectrum is then recorded.

9.4 Chemical shifts

The use of NMR in structural identification relies on the fact that the resonance frequency of the nuclei of a particular isotope varies depending on the chemical environment of the nucleus. This arises from the magnetic shielding of each nucleus by the electrons of the atom with the consequence that each nucleus experiences a magnetic field which is less than the applied

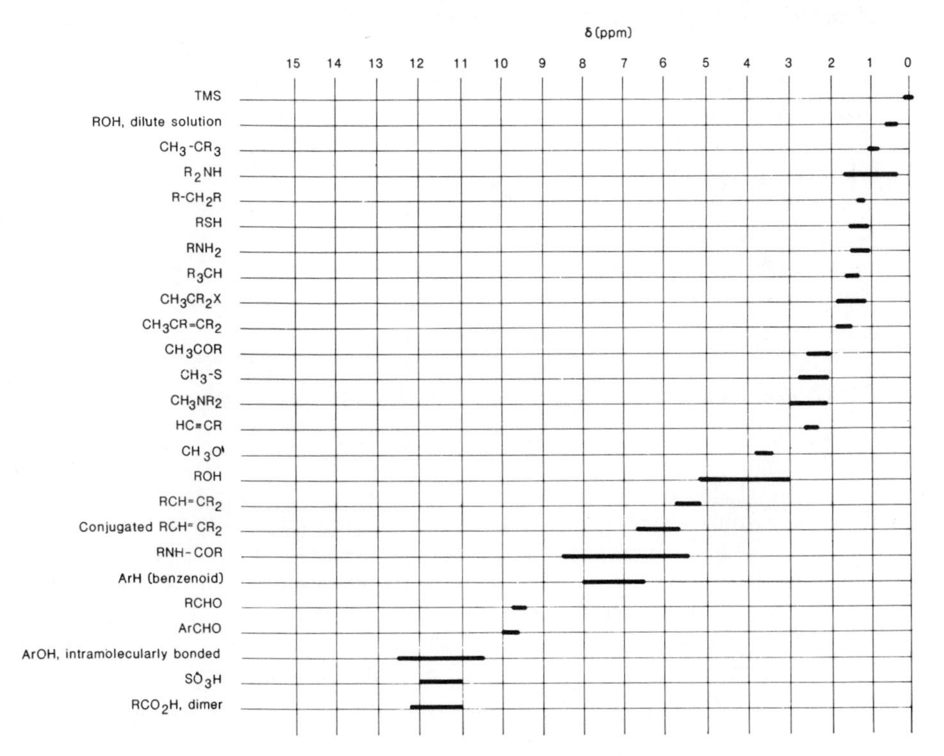

Figure 9.5 Proton chemical shifts. R = saturated alkyl group.

magnetic field B_0. The effective magnetic field at a nucleus is

$$B(\text{nucleus}) = B_0(1 - \sigma)$$

where σ is the shielding or screening constant. It is a dimensionless parameter with units of ppm. The value of σ for each nucleus is not quoted in absolute terms, but the shielding is quoted as the chemical shift δ with respect to a reference standard:

$$\delta = \frac{\nu_s - \nu_R}{\nu_R} \times 10^6 \text{ ppm}$$

where ν_s and ν_R are the resonance frequencies for the sample and reference respectively. A positive value of δ indicates a greater degree of shielding in the reference than in the sample. An increase in shielding corresponds to a reduction in the resonance frequency. Early literature reported chemical shifts in terms of the τ scale, where $\tau = 10 - \delta$. Reference standards include tetramethylsilane (TMS) for ^1H and ^{13}C spectra, and orthophosphoric acid for ^{31}P spectra. TMS contains highly shielded protons and carbon atoms and therefore chemical shifts are generally positive in both ^1H and ^{13}C spectra. Most ^1H chemical shifts are in the range 0–10 ppm (Figure 9.5), while ^{13}C chemical shifts are usually in the range 0–240 ppm (Figure 9.6).

Electronegative atoms and alkenyl or aromatic groupings tend to cause deshielding of nuclei attached to these groupings. The effects of unsaturated groups arise from the circulation of π electrons in a magnetic field (Figure 9.7). This movement of π electrons induces a secondary field. In the case of aromatic compounds and alkenes, the induced field opposes the external magnetic field

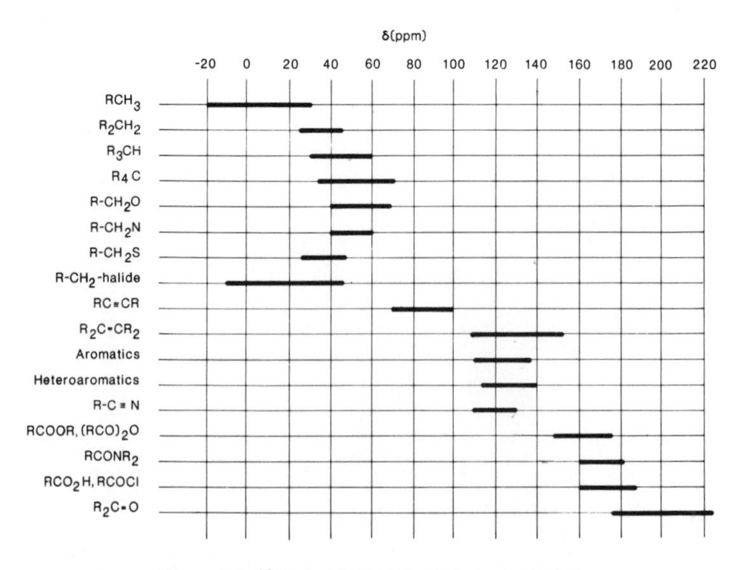

Figure 9.6 ^{13}C chemical shifts (relative to TMS).

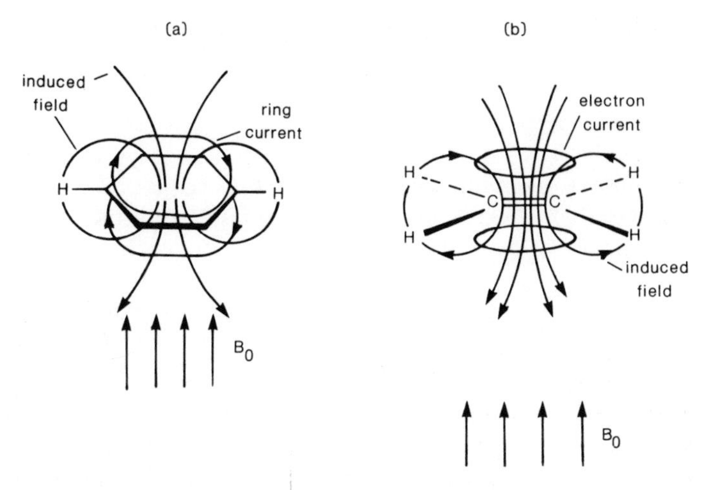

Figure 9.7 (*a*) Deshielding of aromatic protons brought about by the ring current. (*b*) Deshielding of ethylene brought about by the electron current.

and hence causes a deshielding of the protons. However, in an acetylenic bond, the protons are shielded by the induced field when the bond is aligned parallel to the external field, and acetylenic protons have small chemical shifts. The chemical shift facilitates the interpretation of the NMR spectra of complex molecules. For example, the spectrum of penicillin G (Figure 9.8) has proton chemical shift values in the order phenyl $> C_5, C_6 > H_2O > C_3 > C_{10} > C_{17}, C_{18}$. Thus the aromatic protons have the greatest chemical shift, with the C_5 proton, which is deshielded by a neighbouring sulphur and nitrogen atom, and the C_6 proton, which is deshielded by a neighbouring nitrogen atom

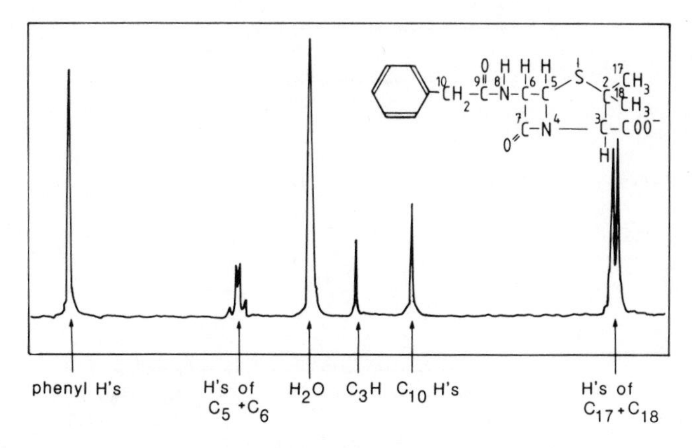

Figure 9.8 Proton NMR spectrum of penicillin G. Redrawn with permission from Fischer and Jardetzky (1965).

and carbonyl group, having the next highest chemical shift. The C_3 and C_{10} protons also are deshielded compared to the high field methyl resonances.

9.5 Spin–spin coupling

The NMR line of a given nucleus can be split into a multiplet by interaction with other nuclei, which is transmitted by the bonding electrons. The number of lines in a multiplet depends on the number of nuclei that the nucleus is

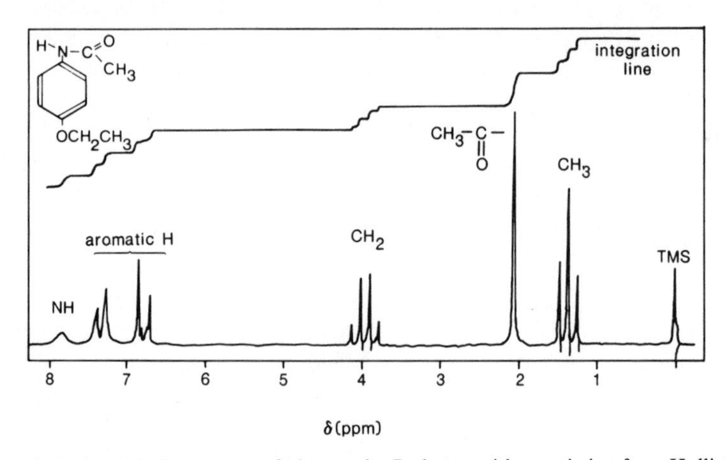

Figure 9.9 Proton NMR spectrum of phenacetin. Redrawn with permission from Hollis (1963).

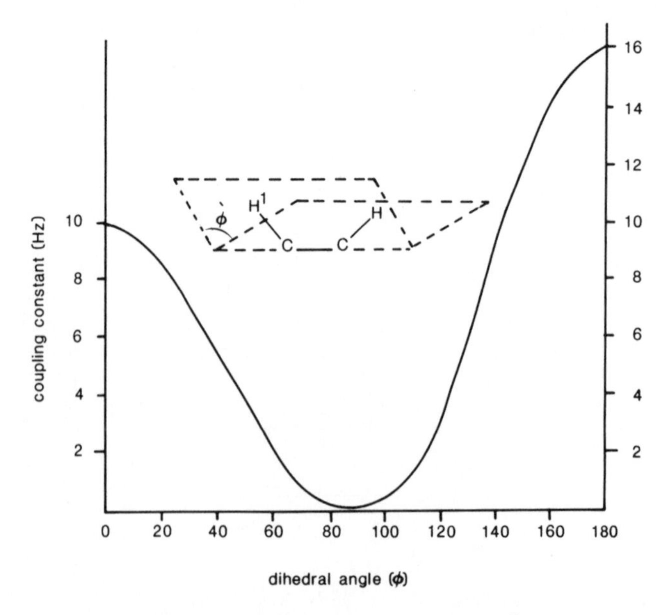

Figure 9.10 Effect of dihedral angle on proton coupling constants.

interacting with. Interaction with n equivalent nuclei yields $n+1$ lines in the multiplet. Equivalent nuclei are nuclei that are in chemically identical environments under the conditions of the experiment. The relative intensities of the lines correspond to the coefficients in a binomial expansion. Thus, one neighbouring proton splits the observed resonance into a doublet (relative intensities 1:1), two produce a triplet (1:2:1), three give a quartet (1:3:3:1) and four produce a quintet (1:4:6:4:1). For example in the ^{1}H spectrum of phenacetin (Figure 9.9), the methyl protons are split into a triplet by the neighbouring methylene group, while the methylene protons are split into a quartet by the methyl group. The $n+1$ lines are equally spaced with the frequency separation between adjacent lines being related to the strength of coupling denoted by J, the coupling constant. In most molecules, weak couplings cause a further splitting of the multiplet arising from the strong coupling. The coupling constant generally decreases with an increase in the

Table 9.2 Proton spin–spin coupling constants

Structure	*J* (*Hz*)	*Structure*	*J* (*Hz*)
		benzene	*o* 6–9 *m* 1–3 *p* 0–1
$>CH-CH<$	2–9	pyridine (ring positions 1 (N)–6)	(2–3) 4·9–5·7 (2–4) 1·6–2·6 (2–5) 0·7–1·1 (2–6) 0·2–0·5 (3–4) 7·2–8·5 (3–5) 1·4–1·9
CH_3-CH_2-X	6·5–7·5		
$(CH_3)_2CH-X$	5·5–7·0	pyrrole (N–H, ring positions 1 (N)–5)	(1–2) 2–3 (1–3) 2–3 (2–3) 2·0–2·6 (3–4) 2·8–4·0 (2–4) 1·5–2·2 (2–5) 1·8–2·3
cyclohexane (H_a, H_e)	a, a 8–10 a, e 2–3 e, e 2–3	thiophene (S)	(2–3) 4·6–5·8 (3–4) 3·0–4·2 (2–4) 1·0–1·8 (2–5) 2·1–3·3
cis $H-C=C-H$	7–12	furan (O)	(2–3) 1·6–2·0 (3–4) 3·2–3·8 (2–4) 0·6–1·0 (2–5) 1·3–1·8
trans $H-C=C-H$	13–18		
$>C=C(H)-C-H$	4–10		

number of chemical bonds between the coupled nuclei, but it is also dependent on the hybridization, dihedral bond angles and electronegativity of substituents. The dependence of proton coupling constants on dihedral angle is shown in Figure 9.10. Table 9.2 indicates the range of values for the coupling constant corresponding to interactions between protons in different environments. The combination of chemical shift data and spin–spin splitting patterns is the basis for the assignment of NMR signals.

9.6 Integration

The area under an absorption band is proportional to the number of nuclei resonating at that frequency. Commercial NMR spectrometers include an electronic integrator which records a continuous line across the spectrum with steps above each peak proportional in height to the area of the peak (see Figure 9.9). This integration line is valuable in structural elucidation and it also makes possible the quantitative analysis of mixtures by NMR. For example, the concentrations of aspirin, phenacetin and caffeine can be determined in mixtures sold as commercial analgesic preparations (Hollis 1963). The reported errors averaged 1·1% for aspirin, 2·2% for phenacetin and 3·2% for caffeine. The sensitivity and accuracy of quantitative determinations by NMR are generally poorer than for other instrumental techniques, such as GLC.

9.7 Further techniques for elucidation of NMR spectra

After studying the chemical shifts, spin–spin splitting pattern, coupling constants and integration, it is often necessary to use further techniques for the identification of an unknown compound having a complex spectrum. Possible techniques include the following.

9.7.1 *Recording the spectrum at higher field strength*

The resonance frequency is proportional to the applied field strength and therefore increasing the applied field strength amplifies the difference in resonance frequency between multiplets, while having no effect on coupling constants. The fine detail of a spectrum is therefore often more evident at higher field strength. Figure 9.11 illustrates the increased separation of multiplets when the ^{1}H spectrum of an azasteroid is recorded at 220 MHz compared with the spectrum at 60 MHz.

9.7.2 *Addition of D_2O*

Water exchanges rapidly with labile protons in functional groups such as —OH, —NH and —COOH, and therefore D_2O will cause the collapse of these signals, since deuterium does not absorb in the ^{1}H spectral frequency range.

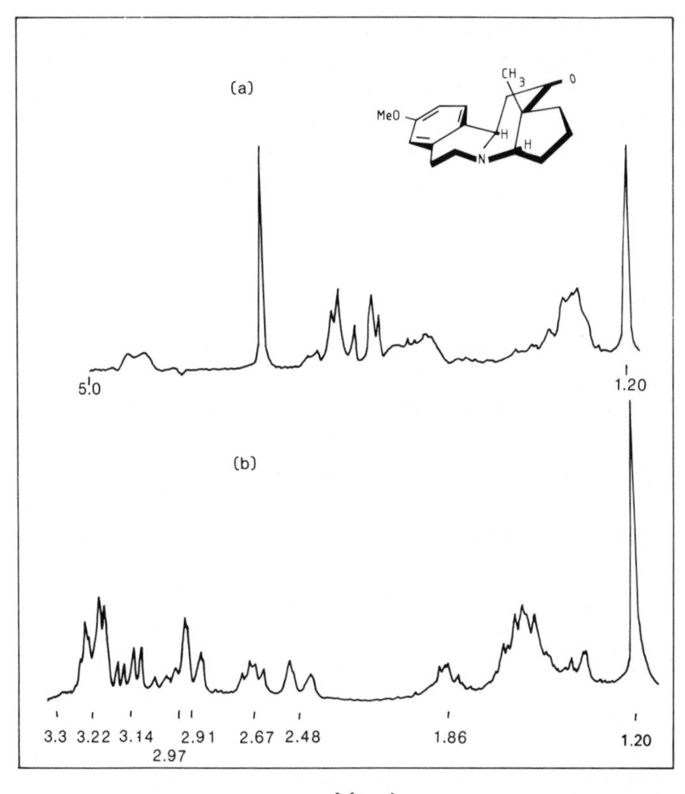

Figure 9.11 Proton NMR spectrum of an azasteroid recorded at (*a*) 60 MHz and (*b*) 220 MHz. Redrawn with permission from Bhacca *et al.* (1968).

9.7.3 *Double-resonance experiments*

These experiments involve irradiating a sample with a second rf field in the plane perpendicular to B_0. Several types of double-resonance experiments can be performed to achieve simplification of the spectrum, to reveal which nuclei are coupled together or to improve sensitivity.

9.7.3.1 *Spin-decoupling.* This technique involves irradiating a nucleus, or group of nuclei, at their resonance frequency. The orientation of the nuclei becomes indeterminate and the spin–spin coupling with the other nuclei is removed, thereby simplifying the spectrum. An important application of this technique is in the recording of ^{13}C spectra.

The couplings between ^{1}H and ^{13}C nuclei can be removed by irradiating the sample with a range of frequencies covering the whole ^{1}H spectrum. This broad-band decoupling simplifies the spectrum considerably. The ^{13}C spectrum of alanine (Figure 9.12) indicates the simple appearance of a ^{13}C

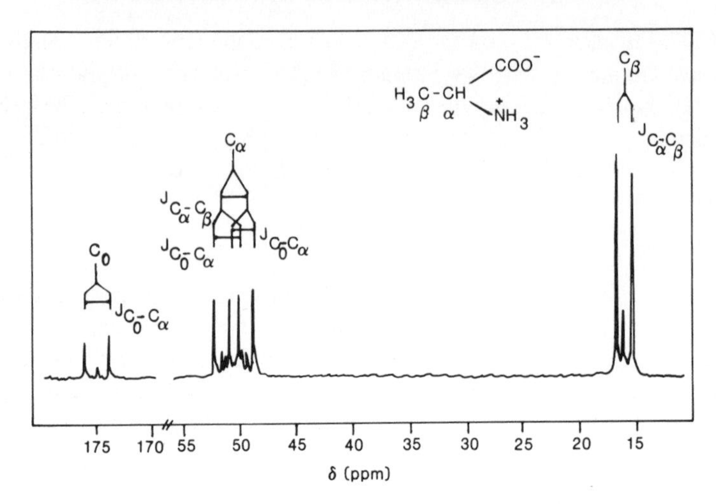

Figure 9.12 ^{13}C NMR spectrum of alanine (J = coupling constant). Redrawn with permission from Tran–Dinh *et al.* (1974).

spectrum recorded with proton decoupling. The carboxylate and methyl carbon atoms are each split into a doublet by C_α, and C_α appears as a quartet due to a larger coupling with the carboxylate carbon atom superimposed on a small splitting with the methyl carbon atom. Coupling constants can readily be derived from the spectrum as shown. Homonuclear spin-decoupling in which spin–spin coupling between nuclei of the same isotope is removed by spin-decoupling is also a useful aid to structural identification. If the applied rf field is weak, complete decoupling does not occur, but additional splitting of spectral lines occurs. The splitting is only symmetrical if the irradiation frequency is exactly equal to the resonance frequency of the other nucleus and this technique, known as spin tickling, is therefore helpful in identifying which nuclei are coupled together.

9.7.3.2 *Internuclear double resonance* (*INDOR*). Spin decoupling involves scanning the spectrum by varying the detection frequency with the decoupling frequency kept constant. In contrast to this, INDOR involves monitoring the signal at a fixed frequency corresponding to a spectral line while the decoupling frequency is varied. When the signal disappears, due to the collapse of a multiplet into a singlet, the decoupling frequency corresponds to that of the coupled nucleus. Hence, the observed spectrum is that of all the nuclei coupled to a particular nucleus, which is simpler and easier to interpret than that of the whole molecule. INDOR allows the resonance frequencies and chemical shifts of various isotopes to be determined with a proton magnetic resonance spectrometer.

9.7.3.3 *The nuclear Overhauser effect.* When a nucleus is irradiated at its resonance frequency with an intensity that causes the populations of the nuclear energy levels to become equal, thus saturating the signal, the intensity of the signals from other nuclei may change. This effect can give information about internuclear distances and the conformation of a molecule.

9.7.4 *Shift reagents*

Paramagnetic ions cause changes in the chemical shift of protons and carbon atoms close to the binding site of the ion. Lanthanide ions can be added to samples to aid identification and assignment of signals. Thus signals in the NMR spectrum that are not separated are shifted to different extents by ions such as europium, and this can allow identification of the multiplets arising from different nuclei.

9.7.5 *Two-dimensional NMR*

Application of a pulse sequence followed by detection of the FID can lead to a separation of the interactions between nuclei. Recording the FIDs, as the time between pulses is varied, can separate coupling information from chemical shift data and this is best represented as a two-dimensional plot. These two-dimensional NMR spectra are particularly useful in unravelling the complexities of the ^{1}H spectra of molecules with a high relative molecular mass. Various experiments of this type can be performed depending on the number of pulses, the power of each pulse (90° or 180°), their precise frequencies, amplitudes, phases and separation in time.

9.8 Wide-line NMR

Wide-line spectrometers operate at low field strengths and do not resolve the absorbances of nuclei in different chemical environments. Wide-line NMR can be used to determine the concentration of a particular isotope in a sample. It also has important applications in studies of the physical state of samples, since the relaxation times are strongly dependent on the degree of molecular mobility. For example, the solid-fat: liquid-fat ratio in an edible fat such as cocoa butter can readily be determined from the FID (Figure 9.13), since the solid-fat signal decays within about 70 μs; the liquid-fat signal decays considerably more slowly, within about 10^4 μs. The solid-fat content determined from the height of the FID at 12 μs must be corrected by a factor which takes into account decay during the period immediately following the pulse. The liquid-fat content is proportional to the height of the FID after 70 μs. Other applications of wide-line NMR include determination of the oil content of oilseeds, moisture content of foods, fluorine content of plastics and studies of free and bound water in biological systems. The principles and selected

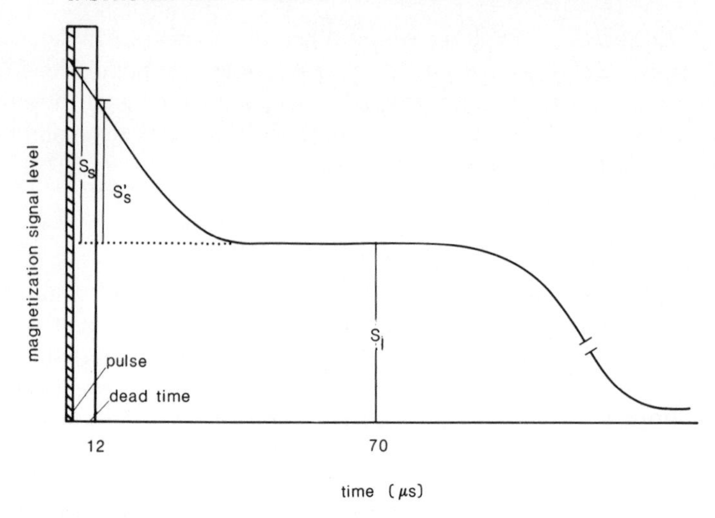

Figure 9.13 Magnetization decay from a fat sample after 90° pulse using a 20 MHz spectrometer.
$\%$ solid phase $= 100\, S_s/(S_s + S_l)$
$= 100 f S'_s/(f S'_s + S_l)$

applications of wide-line NMR are discussed in more detail by Waddington (1986).

9.9 In-vivo NMR

The development of large-bore high-field magnets (up to 1 m at 2 T) has made it possible to perform NMR experiments on intact living animals of substantial size (up to and including man).

In order to study the chemistry of biological samples *in vivo*, it is necessary to separate the signals arising from the nuclei of interest from the background which would be contributed by other organs, tissues and fluids if a classical NMR study was performed, and this may be achieved by a number of techniques. Budinger and Lauterbur (1984) have reviewed recent developments in this area.

9.9.1 *Topical NMR*

One approach to *in vivo* NMR studies involves the shaping of the magnetic field to provide high homogeneity over a relatively small region. High-resolution spectra are obtained from this selected volume and signals from nuclei outside the homogeneous zone are broadened and do not contribute significantly to the spectrum. This technique is known as topical NMR (from the Greek *topos*, a place). A small homogeneous volume can be achieved by a combination of three magnetic field gradients which are modulated sinusoid-

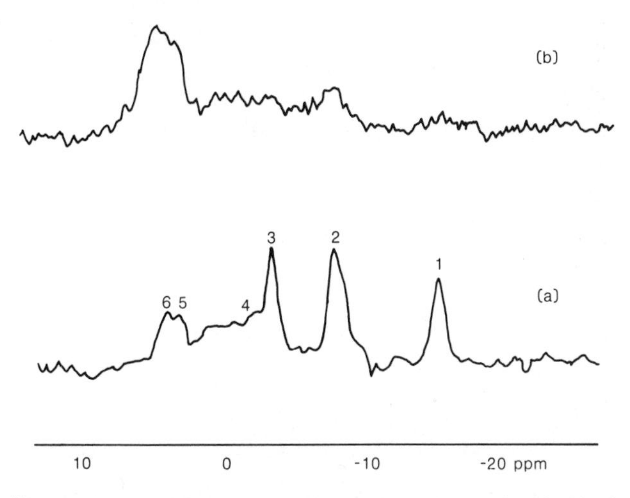

Figure 9.14 ^{31}P–NMR spectra of a live rat anaesthetized with pentabarbitol in the presence of localizing fields designed to record signals from the liver, (*a*) before surgery, (*b*) after surgery to cut off the hepatic blood supply. Peaks 1, 2, 3 are due respectively to the β-, α-, and γ-phosphates of ATP; 4 is phosphocreatine; 5 is inorganic phosphate and 6 is sugar phosphate. Redrawn with permission from Gordon et al. (1980).

ally. Topical NMR has been applied to studies of ^{31}P and ^{13}C. Figure 9.14 shows the ^{31}P NMR spectrum of the liver of a live rat. After surgery to cut off the hepatic blood supply (Figure 9.14*b*), the resonance peaks 1, 2 and 3 corresponding to the β-, α-, and γ-phosphate groups of ATP decrease dramatically, while peak 5 due to inorganic phosphorus increases. Peak 6 is the resonance of sugar phosphate.

9.9.2 *Surface coils*

An alternative method of studying *in vivo* chemistry involves the use of a flattened coil of wire as the NMR probe. The coil is placed on the surface of the animal and detects signals from a cylindrical region extending below the plane of the coil. The rf power required to excite nuclei increases with distance when the coil is used as a transmitter and various procedures have been developed to overcome the consequent limitations on spatial selectivity.

9.9.3 *NMR imaging*

If a linear field gradient is applied to a sample, a signal can be produced in which the magnetization varies with distance. These signals can be combined to form a two- or three-dimensional map or picture which has similarities to that produced by an X-ray CT (computed tomography) scanner. This technique appears to have enormous potential for medicine and it is the subject of considerable research. NMR imaging has been applied to most of the organs of the human body; for example, the technique can provide an accurate non-invasive method of studying cardiac dimensions and function (Longmore *et al.* 1985). Protons have been used in the studies to date, but other nuclei, including ^{13}C, may have applications if the experimental difficulties can be overcome.

References

Bhacca, N.S., Meyers, A.I. and Reine, A.H. (1968) *Tetrahedron Lett.* **19**, 2293.
Budinger, T.F. and Lauterbur, P.C. (1984) *Science* **226**, 288.
Fischer, J.J. and Jardetzky, O. (1965) *J. Amer. Chem. Soc.* **87** (14) 3237.
Gordon, R.E., Hanley, P.E., Shaw, D., Gadian, D.G., Radda, G.K., Styles, P., Bore, P.J. and Chan, L. (1980) *Nature* **287**, 736.
Hollis, D.P. (1963) *Analyt. Chem.* **35** (11), 1682.
Longmore, D.B., Underwood, S.R., Hounsfield, G.N., Bland, C., Poole-Wilson, P.A., Denison, D., Klipstein, R.H., Firmin, D.N., Watanabe, M., Fox, K., Rees, R.S.O., McNeilly, A.M. and Burman, E.D. (1985) *The Lancet*, June 15th, 1360.
Tran-Dinh, S., Fermadjian, S., Sala, E., Mermet-Bouvier, R., Cohen M. and Fromageot, P. (1974) *J. Amer. Chem. Soc.* **96** (5) 1484.
Waddington, D (1986). Applications of wide-line nuclear magnetic resonance in the oils and fats industry, ch. 8, in *Analysis of Oils and Fats*, Hamilton, R.J. and Rossell, J.B. (eds.), Elsevier, London.

Further reading

Becker, E.D. (1980) *High Resolution NMR*, 2nd edn, Academic Press, New York.
Gadian, D.G. (1982) *Nuclear Magnetic Resonance and its Applications to Living Systems*, Clarendon, Oxford.
Gunther, H. (1980) *NMR Spectroscopy – An Introduction*, Wiley, Chichester.
Harris, R.K. (1983) *Nuclear Magnetic Resonance Spectroscopy*, Pitman, Marshfield, MA.
James, T.L. (1975) *Nuclear Magnetic Resonance in Biochemistry*, Academic Press, London.
Jardetzky, O. and Roberts, G.C.K. (1981) *NMR in Molecular Biology*, Academic Press, New York.

10 Electron spin resonance

10.1 Principles

Electron spin resonance (ESR), sometimes called electron paramagnetic resonance (EPR), is a technique for studying the structure and properties of species containing unpaired electrons. Thus it is restricted to free radicals, paramagnetic metal ions and molecules in a triplet electronic state. The technique involves the absorption of microwave radiation, which induces transitions between electronic magnetic energy levels.

Many of the principles of NMR apply also to ESR. In common with many nuclei, the electron has a spin of $\frac{1}{2}$ and an associated magnetic moment. In the applied magnetic field, two energy levels are present corresponding to M_s, the angular momentum quantum number, having values of $+\frac{1}{2}$ and $-\frac{1}{2}$. A high-frequency magnetic field in the plane at right angles to the permanent field induces transitions between the energy levels. Since the electron has a high magnetic moment compared with hydrogen nuclei, the energy difference between the levels, and consequently the radiation frequency, is correspondingly higher. The resonance frequency ν is related to the magnetic field strength B_0 by the equation

$$h\nu_1 = g\beta B_0$$

where h is the Planck constant, β is the Bohr magneton, and g is the Landé g-factor or spectroscopic splitting factor. The Bohr magneton is the magnetic moment of the electron and the spectroscopic splitting factor is a number dependent on the orbital and spin quantum numbers of the particular radical. Generally, a field strength of about 0·33 T is used, corresponding to a resonance frequency of about 9 GHz which is in the microwave region. The larger difference in energy between the ground state and the excited state in ESR compared with NMR leads to a greater difference in population between the spin states at thermal equilibrium and hence increased sensitivity. As little as 10^{-11} mol of a paramagnetic species may be detected by ESR. This sensitivity, however, only applies if the line width is narrow.

Spin relaxation is faster for excited electron-spin states than for the analogous nuclear states, and this leads to broader line widths in ESR. In particular, some transition metal ions including Fe(III) have extremely efficient relaxation processes and the spectral line width is consequently very broad.

10.2 ESR spectra

An ESR spectrum is usually presented as the first derivative rather than the absorption curve. As described for UV spectroscopy (Section 6.10), the zero crossing point can be accurately determined and the derivative mode is particularly suitable for broad poorly-resolved peaks. A typical ESR spectrum in the derivative form is shown together with the absorption spectrum in Figure 10.1. In some cases the second derivative spectrum is plotted. This has the advantage that the peak occurs at the absorption maximum, but in a negative direction.

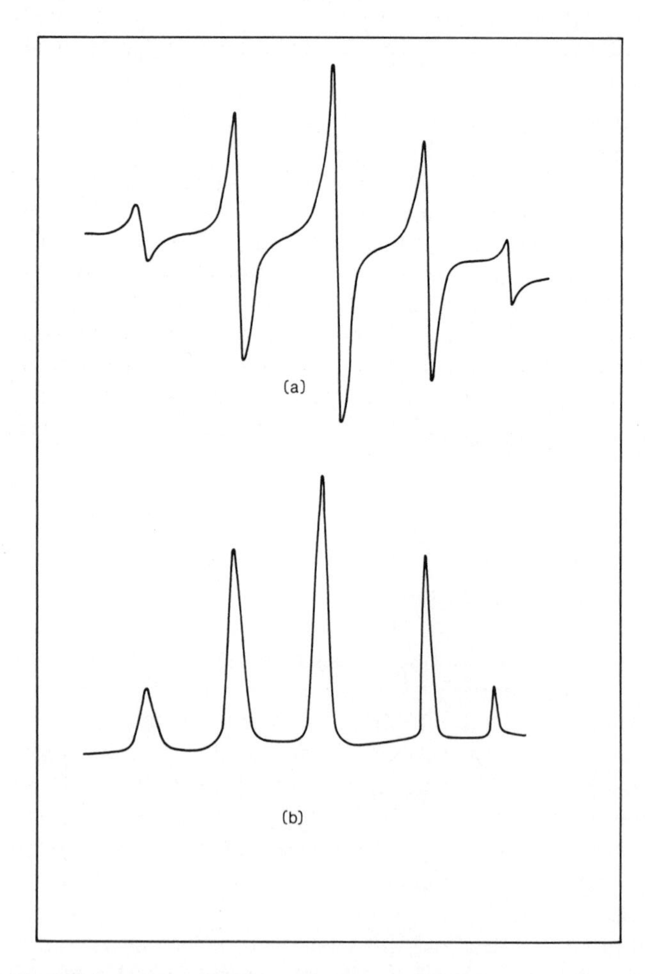

Figure 10.1 The ESR spectrum for the semiquinone radical: (*a*), derivative spectrum; (*b*), absorption spectrum. Redrawn with permission from Venkataraman and Fraenkel (1955).

10.2.1 *The g-factor*

The g-factor represents the constant of proportionality between the field strength and the resonant frequency. It has the value of 2·0023 for a free electron, but organic free radicals have g-factors which vary slightly from this value. Transition metals have values which vary widely between 1 and 18. The g-factor of transition metals is also often strongly anisotropic, varying according to the orientation of the molecule to the applied magnetic field. This leads to broad and asymmetric ESR peaks. The g-factor can provide information about associated ligands and their orientation in the region of paramagnetic metals, but is not generally useful for organic free radicals. The limited variability of the g-factor for organic radicals is in contrast to the chemical shift in NMR, which provides much valuable structural information.

10.2.2 *Hyperfine splitting*

The ESR signal is subject to hyperfine splitting due to the interaction of the electron magnetic moment with nuclear magnetic moments. This is analogous to spin–spin coupling in NMR. An electron interacts with atoms such as ^{1}H, ^{13}C, ^{14}N that have a nuclear spin. In general, interaction with n equivalent nuclei of spin $\frac{1}{2}$ yields $n + 1$ signals with relative intensities according to the coefficients of the binomial expansion. Non-equivalent nuclei yield 2^n lines.

Hyperfine splitting patterns are valuable for the identification of radicals. This is illustrated in the ESR spectrum of α-tocopheroxyl radicals (Figure 10.2). The hyperfine splitting by the six protons on the two neighbouring methyl groups produces a pattern of seven lines.

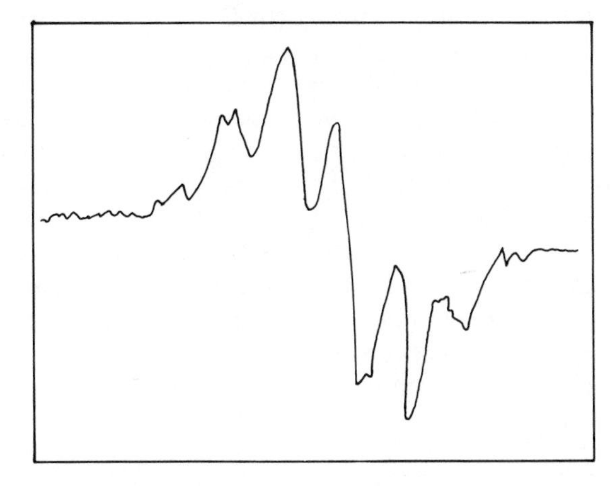

Figure 10.2 First derivative ESR spectrum of α-tocopheroxyl radicals in di-*t*-butyl peroxide and hexadecane solution obtained by photolysis at 320 K. Redrawn with permission from Bascetta *et al.* (1983).

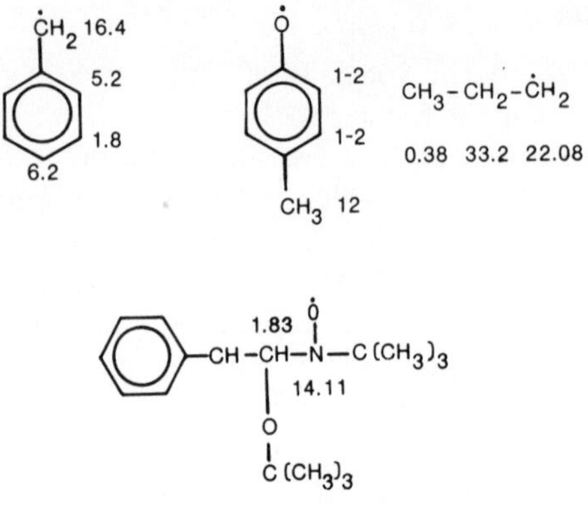

Figure 10.3 Hyperfine splitting constants (in gauss).

An electron localized on an atom shows strong coupling with an α or β hydrogen atom, but couplings with more remote atoms are weak or not detectable. Double bonds or aromatic rings extend the range of strong coupling. The extent of coupling is shown by the hyperfine splitting constant, which is the separation of two adjacent peaks in a spectrum arising from coupling with a particular atom, or group of equivalent atoms. Some typical values are shown in Figure 10.3.

Electron–electron spin coupling can occur in species containing two unpaired electrons, such as molecules in the triplet state. This fine splitting is generally an order of magnitude stronger than hyperfine splitting.

Double-resonance techniques including electron nuclear double resonance (ENDOR) or electron double resonance (ELDOR) may be used for sharpening resonance lines or separating overlapping radical spectra. ENDOR involves irradiation of the sample with a microwave frequency suitable for electron resonance and an rf suitable for nuclear resonance. The ESR spectrum is observed at one frequency while the rf is swept. The variation of signal height with rf allows identification of the nuclear transitions responsible for broadening the ESR line.

ELDOR involves irradiation of a sample with two microwave frequencies. One of these causes electron resonance at a particular frequency, while the other can be swept through other regions of the spectrum. The ESR signal height is recorded as a function of the difference in the two microwave frequencies. This technique can separate signals arising from more than one radical.

10.3 ESR spectrometer

ESR spectra are obtained by scanning the magnetic field rather than the frequency. Hence the major components of an ESR spectrometer (Figure 10.4)

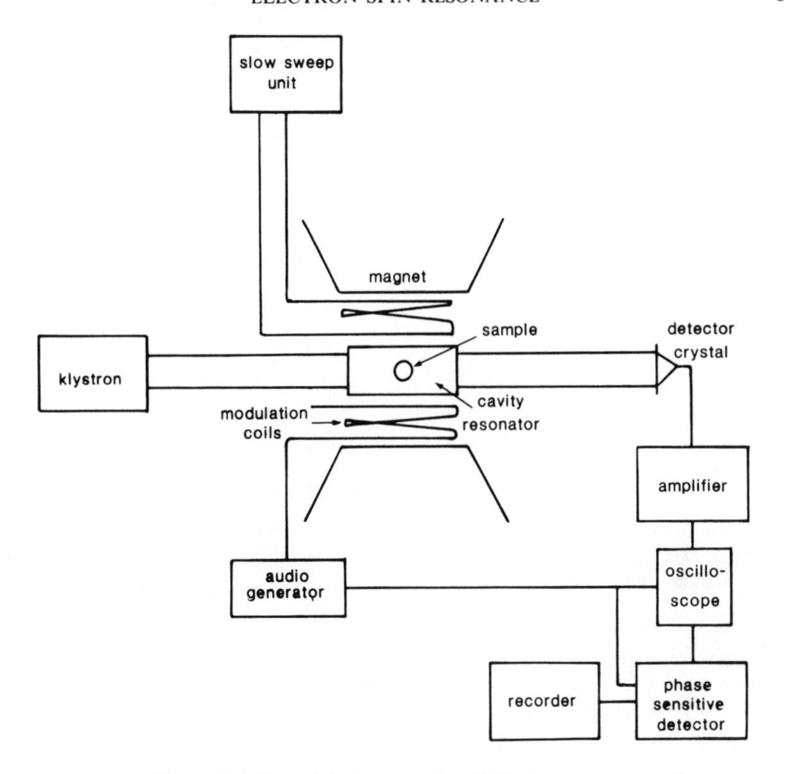

Figure 10.4 Essential elements of an ESR spectrometer.

include a source of microwave radiation of fixed frequency, a constant homogeneous magnetic field and a method of varying the magnetic field to allow the spectrum to be scanned. A detector is used to measure the microwave power absorbed, and the spectrum can be recorded on a chart recorder.

A Klystron, which is a special type of valve oscillator, is used as a source of monochromatic microwave radiation. The radiation is led down a rectangular copper pipe, or waveguide, to the sample cavity in the magnetic field. In order to optimize the sensitivity of detection, a bridge technique is used. The waveguide has a 'magic -T' configuration as shown in Figure 10.5. The arms of the T are linked to the sample, the detector and an adjustable load or absorber of microwave energy. The load is adjusted to absorb the same amount of energy as the sample at a field away from resonance. The spectrum is scanned by altering the magnetic field with sweep coils mounted on the poles of the magnet. When the sample absorbs more energy at the resonant field strength, the bridge becomes unbalanced and microwave power flows towards the detector. A semiconducting silicon–tungsten crystal, which converts the microwave power into a direct current signal, is used as the detector. The applied field is modulated at a frequency of 100 kHz. This improves the signal-

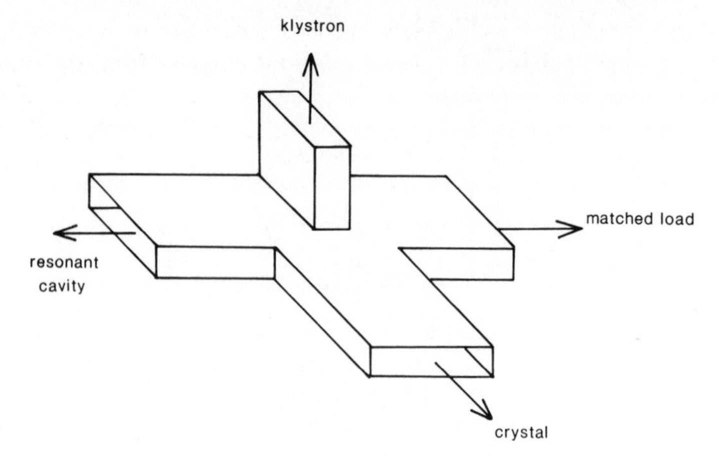

Figure 10.5 Waveguide for an ESR spectrometer.

to-noise ratio. The spectrum is produced as the first derivative of the absorption curve.

10.4 Sample preparation

Solid, liquid or gaseous samples can be used in an ESR spectrometer. Since virtually all organic free radicals are unstable, they must be prepared shortly before the spectrum is scanned. Oxidation or reduction of stable molecules, or irradiation of samples with intense UV, X-ray or γ-radiation may be used to generate radicals. Sample preparation may include freezing a sample into a glassy matrix before irradiation, rapid freezing of samples after reaction, use of a continuous-flow system, and operating under nitrogen. These procedures are used to maintain radical concentrations for a sufficient time to record a spectrum. Samples of biological interest are often prepared as aqueous solutions and frozen before the spectrum is run.

Paramagnetic metals in biological systems give broad peaks resulting in poor resolution and low sensitivity. Freezing in liquid nitrogen can improve the sensitivity by increasing the population of the ground state.

10.5 Spin labelling

Stable free radicals or spin labels may be chemically bonded to biological molecules. This technique converts a molecule that cannot be investigated by ESR to a radical suitable for study and it has increased the scope of ESR considerably. Stable nitroxide radicals ($R_2N—\dot{O}$) are often used as spin labels. A sharp, simple ESR spectrum is often produced, which may give useful information about the chemical environment of the label. The technique has considerable potential for studies of macromolecules, including proteins and nucleic acids. Spin labels with specific chemical structures react with selected functional groups in biological molecules, and the use of various labels may be valuable in gaining information about different parts of the molecule. However, care is required in using spin labels since they may introduce conformational or other significant changes into the molecule. An example of spin labelling was reported by Jones *et al.* (1972) in studies on phosphofructokinase. This enzyme is involved in the regulation of glycolysis and contains one highly reactive sulphydryl group per subunit. The common spin label 4-(2-iodoacetamido)-2, 2, 6, 6-tetramethylpiperidinooxyl was used.

CH_3 CH_3
I—CH_2CONH— NȮ
CH_3 CH_3

The resultant ESR spectrum (Figure 10.6) was used to gain information about the rotational mobility of the enzymes. Changes during the titration of the enzyme with MgATP could also be monitored by recording the spectrum after each addition of substrate. The modified enzyme retained its normal activity and allosteric kinetic behaviour.

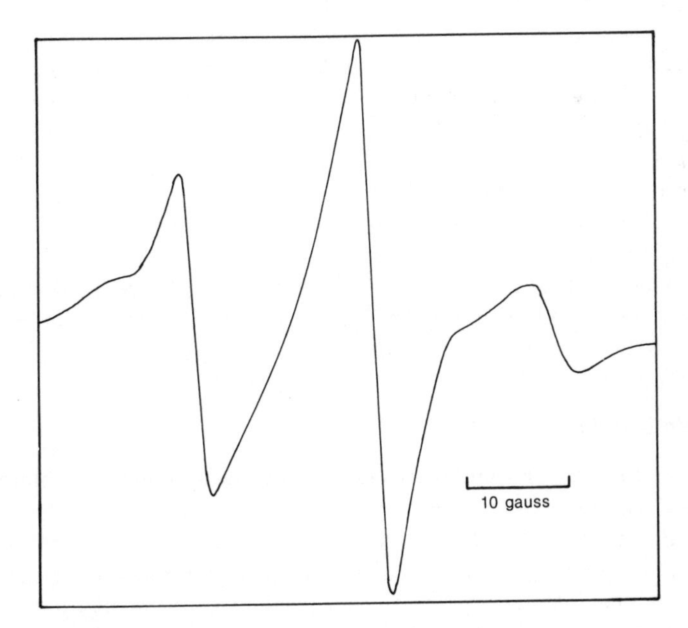

Figure 10.6 First derivative ESR spectrum of spin–labelled phosphofructokinase at pH 7.5. Redrawn with permission from Jones *et al.* (1972).

10.6 Quantitative analysis

The total area under an absorption or derivative ESR peak is proportional to the radical concentration. Often peak-to-peak height measurements are used as a measure of concentration because the signals tail badly, making area measurements difficult. ESR is commonly used for kinetic or other quantitative studies in biochemistry and related areas.

References

Bascetta, E., Gunstone, F.D., Walton, J.C. (1983) *Chem. and Phys. of Lipids* **33**, 207.
Jones, R., Dwek, R.A. and Walker, I.O. (1972) *FEBS Lett.* **26**, 92.
Venkataraman, B. and Fraenkel, G.K. (1955) *J. Amer. Chem. Soc.* **77**, 2707.

Further reading

Knowles, P.F., Marsh, D. and Rattle, H.W.E. (1976) *Magnetic Resonance of Biomolecules*, Wiley, London.
Symons, M.C.R. (1978) *Chemical and Biochemical Aspects of Electron-Spin Resonance Spectroscopy*, Van Nostrand Reinhold, New York.

11 Flame techniques

11.1 Introduction

There are three related techniques, often collectively known as flame techniques, based on the ability of flames to produce free atoms and other atomic species from inorganic analytes in solution. The simplest of these is flame photometry, now more commonly termed flame emission spectrometry (FES), which is based on the measurement of the light emitted when atoms raised to an excited state by the thermal energy of the flame return to the ground state. In practice, the proportion of the atoms raised to an excited state is very small, so that most of the free atoms remain in the ground state. These atoms are thus able to absorb radiation, provided its frequency, and hence energy, corresponds to the difference in energy between the ground and excited states. This process of absorption is the basis of atomic absorption spectrometry (AAS). The absorbed energy will be subsequently re-emitted as the atoms return to the ground state. This re-emission will be in all directions and hence a net absorption of the radiation will occur along the axis of the incident beam. This re-emission also provides the basis for the third flame technique, that of atomic fluorescence spectrometry (AFS), in which the fluorescence is measured at an angle, often 90°, to the incident radiation. This is a relatively specialized technique, but the other two techniques are very widely used in the analysis of biological materials. The essential differences between these three techniques can be summarized diagrammatically (Figure 11.1). It should be noted that in FES, no external radiation source is involved. The nature of the events taking place as the analyte solution enters the flame are of crucial importance to all of these techniques. The conversion of the analyte in solution to a form capable of determination by FES or AAS is a multi-stage process as shown in Figure 11.2. The proportion of species converted at each stage will influence the final yield of the required atomic state in the flame.

The first stage, nebulization, is achieved by drawing the analyte solution into the burner via a narrow capillary as a result of the Venturi action of the high-velocity gas stream across the tip of the capillary. The nebulization may take place away from the burner head (pre-mix burner) or the solution capillary may extend into the base of the flame (total combustion burner). The latter is more efficient, in that it allows the entire solution stream to enter the combustion zone, but is restricted to flames of short path length and so is used mainly for emission applications. The droplet-size distribution of the aerosol, which is important for the subsequent states, is affected by the physical properties of the solution, in particular viscosity, density and surface tension,

FLAME EMISSION SPECTROMETRY (FES)

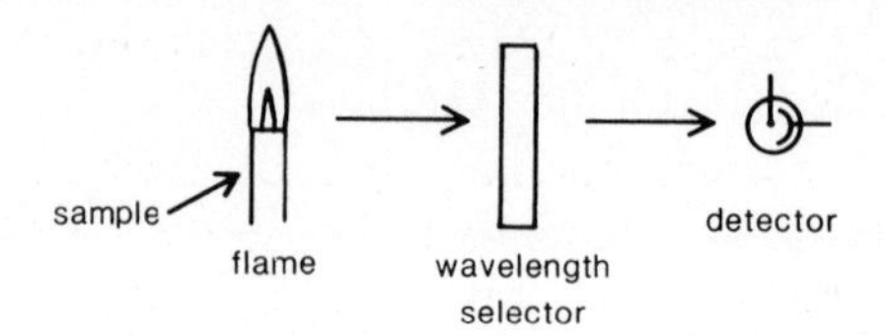

ATOMIC ABSORPTION SPECTROMETRY (AAS)

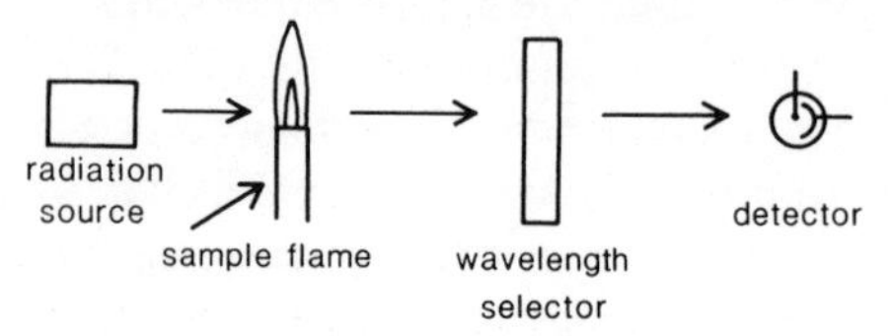

ATOMIC FLUORESCENCE SPECTROMETRY (AFS)

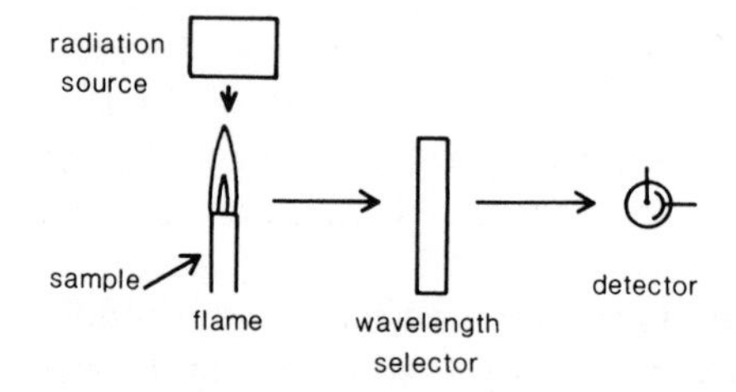

Figure 11.1 Essential features of flame techniques.

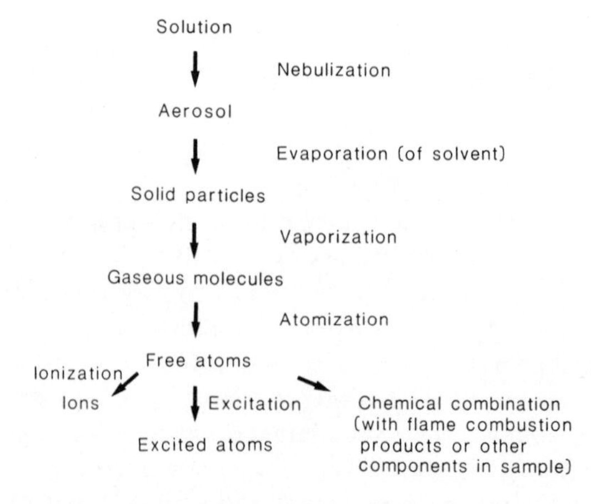

Figure 11.2 Processes involved in conversion of ions in solution to various atomic species within a flame.

as well as the flow rates of the nebulizer gas and sample solution. In practical terms, the most severe limitation is for viscous solutions, such as blood and urine, which would have to be diluted or alternatively the organic material destroyed by wet oxidation.

Provided that the aerosol contains only small droplets, and that the rate of aspiration is not too great, evaporation takes place rapidly at the base of the flame to form solid particles. Water is by far the most commonly used solvent, but in those cases where flammable organic solvents are used, for example, 4-methylpentan-2-one, these will burn in the flame, affecting its characteristics and also possibly contributing to background emission.

The efficiency of the vaporization stage depends on the size of the solid particles formed and the temperature of the flame, together with the residence time of the particles within the flame. The flame characteristics also affect the subsequent stage of atomization. A wide range of fuel/oxidant combinations have been used in flame techniques and the more common ones are shown in Table 11.1. The wide range of temperatures available is important as different elements have different optimum temperatures for atomization, and indeed for excitation. The nature of the flame can also be changed by altering the fuel/oxidant ratio. Thus, a fuel-rich flame will provide a reducing atmosphere that will diminish the tendency for some atoms to form refractory oxides, which would otherwise reduce the number of free atoms available for excitation.

The flame temperature is also critical in terms of the excitation step. Once the free atoms are formed they may be converted to excited states by promotion of electrons to higher energy levels by additional thermal energy or by the absorption of radiation of the correct frequency. Thus, if E_0 represents the ground state energy and $E_1, E_2, \ldots$ excited states, the energy difference may be related to radiation frequency by

$$\Delta E = E_1 - E_0 = h\nu$$

where h is the Planck constant and ν is the frequency of radiation that is

Table 11.1 Common premixed flames

Fuel	*Oxidant*	*Temperature** (°C)
Acetylene	Air	2400
Acetylene	Nitrous oxide	2800
Acetylene	Oxygen	3140
Hydrogen	Air	2045
Hydrogen	Nitrous oxide	2690
Hydrogen	Oxygen	2660
Propane	Air	1925
Natural gas	Air	1700–1900

*Stoichiometric mixture

Table 11.2 Variation of population ratio (N_1/N_0) with wavelength and temperature

Element	*Wavelength* (nm)	N_1/N_0 2000 K	4000 K
Sodium	589	$9{\cdot}86 \times 10^{-6}$	$4{\cdot}44 \times 10^{-3}$
Calcium	423	$1{\cdot}21 \times 10^{-7}$	$6{\cdot}03 \times 10^{-4}$
Zinc	214	$7{\cdot}31 \times 10^{-15}$	$1{\cdot}48 \times 10^{-7}$

required to promote ground-state atoms to the excited state E_1 (AAS) or the frequency of the radiation that would be emitted when excited atoms revert to the ground state (FES). The relative population of the ground and excited states is central to these flame techniques. This population ratio (N_1/N_0), where N_1 and N_0 represent the number of atoms in the respective states, is related to the excitation energy ΔE and the absolute temperature T by the Boltzmann equation

$$N_1/N_0 = K \exp(-\Delta E/kT)$$

where K is a statistical factor and k the Boltzmann constant.

Thus, the higher the temperature the greater the population of the excited state, and also the smaller the excitation energy the greater the excited population. These effects are illustrated for three common elements in Table 11.2. Even under favourable conditions for excitation, small ΔE and high T, the majority of atoms will be in the ground state and so able to absorb radiation, while relatively few will be in the excited state able to emit radiation. This factor is important in terms of the relative sensitivities of FES and AAS which depend on excited-state and ground-state populations respectively. It is also clear that in order to obtain reproducible data from any flame technique, the flame characteristics, including temperature, must be carefully controlled.

The lowest excited state E_1 will be the most populated of the excited states with E_2, E_3,... being populated to progressively lower extents. It is also possible that an atom will receive sufficient thermal energy for its most loosely bound electron to be removed completely, leading to ionization. However, this only takes place to a small extent, even for elements with low ionization energies, such as the alkali metals. Nonetheless, it is a process which reduces the population of free atoms, affecting FES and AAS.

The population of free atoms will also be affected by any chemical combination which may take place with other components in the sample. This type of interference is a severe limitation of flame techniques, but may be overcome in many cases (see Section 11.2.1).

11.2 Flame emission spectrometry (FES)

The major components of a flame emission spectrometer are shown in Figure 11.3. The nebulizer/burner may be of either the premix or complete-

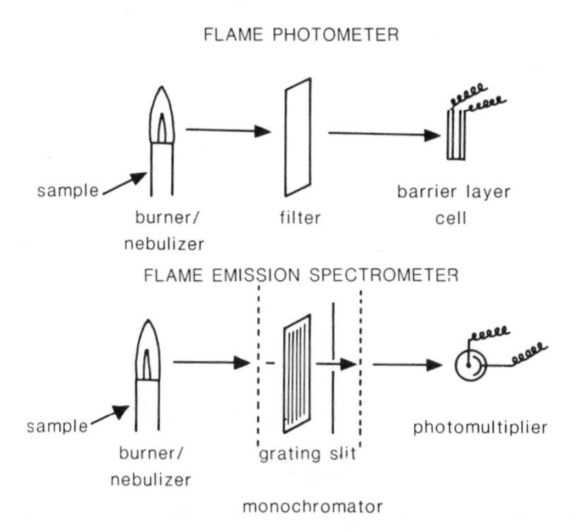

Figure 11.3 Equipment for flame emission spectrometry.

combustion type. In the simplest of instruments, known as flame photometers, the wavelength selection is achieved by an interference filter which has a high level of transmission but a relatively wide bandpass (10–15 nm) and detection of emitted light is achieved with a barrier.layer cell. The poor sensitivity of this type of detector is not a problem in many applications of flame photometry, for example the alkali metals, due to the high levels of radiation emitted, as these analytes are often present in biological materials at high levels. Flame photometers often operate at low flame temperatures (natural gas/air) and are restricted to those elements with low excitation energies, such as sodium, potassium, calcium and lithium.

In flame emission spectrometers, the wavelength selection is achieved by a monochromator, most commonly a diffraction grating, which provides much improved spectral resolution. The detection system is usually a photomultiplier providing greater sensitivity than the barrier layer cell used in flame photometers. The similarity between flame emission spectrometers and atomic-absorption spectrometers will soon become evident, and indeed it is often possible to use an atomic-absorption spectrometer for emission work, that is, with the radiation source turned off. However, for absorption work, a flame of long path length is used and this could lead to problems of self-absorption with emission studies. Self-absorption is a process whereby the light emitted from an excited atom returning to the ground state is absorbed by the surrounding, and much higher, concentration of ground-state atoms, thereby reducing the amount of light emitted from the flame. This effect is only observed with high concentrations of analyte ions in solution and can be reduced by changing the orientation of the burner to the optical axis, so that the light emitted from excited atoms has only to traverse the width, rather than the length of the flame.

Self-absorption may lead to non-linear calibration, with a negative deviation at higher concentrations, where the effect is more pronounced.

FES is subject to a number of interferences, the severity of which depends in some cases on the instrumentation used; some of these interferences will also be encountered in AAS.

11.2.1 *Interference effects*

These effects have been variously classified, but the broad classification into spectral and chemical interferences would seem to be the most sensible.

11.2.1.1 *Spectral interference.* This arises simply from the presence of other species in the flame emitting radiation at a similar wavelength to that of the analyte of interest. If the interfering wavelength is discrete, then it may be possible to remove the interference altogether by increasing the resolution of the wavelength selector, for example by changing from an interference filter to a monochromator. For example, if the interfering species emitted radiation at a wavelength 5 nm away from that of the analyte, this would not be adequately removed by an interference filter (bandpass 10–15 nm), but with a monochromator (bandpass 1–2 nm) no interference at all would be exhibited. On the other hand, if the interfering species emits radiation over a wide range of wavelengths, including that of the analyte, increasing the degree of resolution will reduce the interference but will not eliminate it. Such interference may result from combustion products of the fuel gas, organic material from the sample matrix or species formed from radicals in the flame, for example, CaOH. Such background emission may be removed by taking emission readings either side of the wavelengths of interest. These should be as close as possible to the wavelength of the analyte, taking into account the bandpass of the wavelength selector used. The process involved is illustrated in Figure 11.4, the emission due to the analyte of interest being given by the difference between the emission at A minus the mean of the readings at B and C. Such a

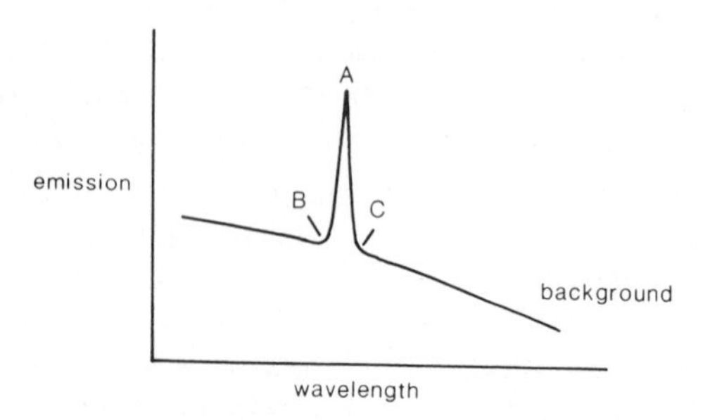

Figure 11.4 Background emission correction.

procedure would only be possible in instruments where precise control of the emission wavelength could be achieved, that is, in monochromator-based instruments as opposed to filter photometers.

11.2.1.2 *Chemical interference.* This type of interference results from incomplete dissociation of the analyte within the flame, leading to a reduction in the formation of free atoms. It may be caused simply by the fact that the analyte is present in solution in a form that does not readily dissociate under the flame conditions used, or that the analyte combines with other ions or radicals within the flame to form refractory compounds. An example of the former type of complication would be the low values recorded for calcium when high levels of sulphate or phosphate are present in solution, due to the formation of the stable compounds calcium sulphate or phosphate. The formation of compounds within the flame is mainly restricted to those atoms that can combine with radicals from the combustion process yielding oxides or species such as CaOH. In most cases, these oxides rapidly break down yielding free atoms, but for certain elements, such as aluminium, vanadium or titanium, the oxides are extremely stable and hence reduce the number of free metal atoms formed.

The extent of these interference effects may be changed by altering the properties of the flame. Thus, if the flame temperature is increased by, for example, changing from natural gas/air to acetylene/air, the degree of dissociation of these more stable compounds will be increased and hence the population of free atoms. Alternatively, if the stoichiometry of the flame is altered to make it 'fuel-rich' and therefore reducing, there will be less tendency for the refractory oxides to be formed. For example, the use of a 'fuel-rich' acetylene/nitrous oxide flame, which is both reducing and provides a high temperature, is required for the determination of aluminium.

An alternative method to overcome these problems of chemical interference is the use of releasing agents, which may be of two types. In the first of these, a high level of a competing ion is added to the test solution so that any interfering anion is effectively removed. In the example cited above, where phosphate is interfering with calcium determinations, a high level of strontium could be added which would preferentially combine with the phosphate due to its high concentration, thus leaving the calcium ions in a free state suitable for atomization. The second approach is to use a protective releasing agent which will combine with the analyte, calcium in our example, preventing reaction with the phosphate. Clearly, any such reagent must be rapidly degraded in the flame to allow free atoms to be formed. Ethylenediaminetetraacetic acid is often used in this role.

In extreme cases where the above methods are not satisfactory, it may prove necessary either to extract the analyte from solution, for subsequent measurement, or to extract the interferent. This may involve the use of complexing reagents to extract the metallic ions of interest into organic solvents or the use of ion-exchange resins.

11.2.1.3 *Ionization interference.* Ionization effects are often categorized as chemical interference but they are not directly caused by chemical interactions. In the flame, a small proportion of excited atoms may ionize, which will reduce the population of excited atoms and hence reduce the intensity of any subsequent emission. This will also affect the population of ground-state atoms and so will have implications for atomic absorption as well. The extent of ionization may be controlled by using the lowest flame temperature feasible, bearing in mind the requirements for dissociation and excitation, that is, relatively high temperatures. The ionization process will also be affected by the presence of other metal ions, in particular those with low ionization energies which will yield a 'pool' of electrons, tending to suppress ionization of the analyte. This effect can therefore give rise to interference, for example, the ionization of calcium is suppressed by the presence of high concentrations of potassium ions, which results in an *increase* in the light emitted by calcium. This also means that ionization can be effectively eliminated by adding a high concentration of such ions which would then act as an ionization suppressant.

11.2.2 *Quantitative measurements*

FES is more commonly used to determine the amount of a particular element in solution, rather than as a qualitative method to show which elements are actually present. In outline, the basis of quantification is simple; determination of the light emitted from the sample under specified instrumental conditions and comparison with a standard solution of the element in question measured under the same conditions. However, the interferences discussed above mean that such simple comparisons may not be adequate and therefore alternative techniques must be employed. Those commonly encountered methods will be discussed in turn, and it should be noted that these same methods find application in many other quantitative techniques.

11.2.2.1 *External standard.* This is the simplest method and involves the construction of a calibration curve covering the concentration range expected for the samples being analysed. The spectrometer would be set for the appropriate emission wavelength (or a filter selected in a photometer) and then zeroed with a blank solution (often ion-free water). The highest concentration standard would then be studied and the sensitivity adjusted to provide a high scale reading, for example, 80% full scale. Intermediate standards would then be analysed under *the same* instrumental settings. A graph of these readings versus concentration will then establish whether the calibration is linear, or at least show the linear range of the instrument. The samples are then analysed, again under the same conditions, and the concentrations read from the graph. If the samples fall outside the linear range, ideally they should be diluted and the analysis repeated, although non-linear calibrations can be used. Clearly, in this simple method, the assumption has been made that no significant interference is taking place and therefore that the analyte in the sample is in exactly the same environment as in the aqueous standards. This situation is rarely encountered in biological samples and great care must be taken in the interpretation of such data. One way to overcome this problem is to make up the calibration standards in a 'synthetic sample matrix' in an attempt to arrange for the analyte in the standards and in the samples to be subject to the same levels of interference. This is never easy and, at best, will only reduce the problem, although it has been successfully used with samples of fairly constant composition, such as milk. A logical extension to this method is to use the sample itself as the matrix for making up the standards, and this is the basis of the next method.

11.2.2.2 *Standard additions.* This method is widely used and should be considered as standard practice, especially where the extent of any possible interference is not known. Here, the light emitted from the sample solution is compared with that from a similar solution to which a known addition of the analyte has been made. If R_{Sa} is the reading from the sample which contains x units (for example, p.p.m.) of the analyte, then

$$R_{\mathrm{Sa}} \propto x$$
$$= kx$$

or

$$k = R_{\mathrm{Sa}}/x$$

Similarly, for the standard (spiked sample) with reading R_{St} and level $(x + a)$ of analyte, where a is the amount added,

$$R_{\mathrm{St}} \propto (x + a)$$
$$= k(x + a)$$

assuming, of course, that the proportionality is the same in both cases (which it would be provided the instrumental sensitivity had not changed). The proportionality constant k can then be eliminated from these equations:

$$R_{\mathrm{St}} = (R_{\mathrm{Sa}}/x)(x + a)$$

$$x = \left(\frac{R_{\mathrm{Sa}}}{R_{\mathrm{St}} - R_{\mathrm{Sa}}}\right)a$$

This assumes that the instrumental response is linear over the working concentration range and this should be confirmed by making a series of additions at different levels to produce a graph as shown in Figure 11.5. In this manner the natural level of the analyte, x, can be determined graphically as shown in the same figure. In this method, if the analyte is subject to interference in the sample, then the same interference will affect the calibration and hence will be compensated. If only a single point addition is made, this will

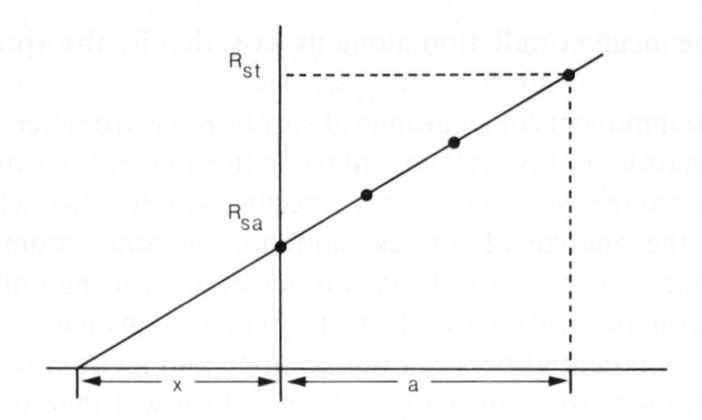

Figure 11.5 Method of standard additions.

automatically assume a linear response, and where this is not the case, errors will be introduced. It should be noted that the method still relies on constant instrumental sensitivity and if this is in doubt further techniques are necessary.

11.2.2.3 *Internal standard method.* The simplest way to overcome problems associated with changes in instrumental sensitivity is to add an internal standard to both the calibration standards and the samples. This should be a metallic ion very similar to the analyte of interest, which is not naturally present and which will be affected by instrumental changes in the same way as the analyte. The same concentration of internal standard is usually added to both the standards and the samples, as this facilitates the calculations, but is not essential. The ratio of the reading for the analyte and the internal standard in the standard solution is then compared with the same ratio for the sample, and hence the concentration of the analyte in the sample determined by direct proportionality. Here again, a linear response has been assumed and a complete calibration curve should be constructed with standard solutions containing increasing amounts of the analyte, but the same concentration of internal standard in each case.

11.3 Atomic-absorption spectrometry (AAS)

AAS is similar to FES in the sense that it relies on the high temperatures within a flame to convert metallic ions in solution to a form suitable for analysis (see Section 11.1). However, there is a fundamental difference in that AAS measures the amount of radiation *absorbed* by the ground-state atoms, rather than the radiation emitted by excited states returning to this ground state, as in FES. A prerequisite for AAS is therefore an external source of radiation of a frequency which corresponds to the transition of ground-state atoms of the analyte of interest to an excited state. This absorbed energy will be subsequently re-emitted as the atoms once more return to the ground state, but, as this will take place in all directions, there will be a net reduction of the intensity of the incident radiation along its axis, that is, absorption will take place.

The major components of an atomic absorption spectrometer are shown in Figure 11.6. In order to take full advantage of the inherent specificity of AAS, the radiation source must emit very specific wavelengths which will be absorbed by the analyte of interest and not by other atoms of similar absorption characteristics as well. This is achieved by using hollow cathode lamps, which contain a cathode of the metal of interest in an inert atmosphere. Application of a potential between this cathode and an adjacent anode will lead to ionization of some of the gas atoms, which will then be accelerated towards the cathode, dislodging metal atoms in an excited state. On returning to the ground state, these atoms emit radiation of a specific wavelength characteristic of the metal of the cathode. Such lamps have a finite life due to vaporization of the cathode material and also loss of the inert gas by absorption on the lamp walls. Those lamps operating in the UV region must contain a silica window which is transparent at these wavelengths. Usually the cathode material is of a single metal, so that a different lamp is required for each element to be analysed. However, it is possible to construct multi-element lamps in which the cathode is an alloy of the metals of interest. The resulting radiation will then contain frequencies corresponding to all these elements and can therefore be used for the appropriate analysis, simply by altering the monochromator setting to allow only specific wavelengths to pass to the detector. The combination of metals that can be used in this manner is restricted by spectral overlap and also by the fact that they must form a stable alloy.

An alternative set of lamps is available which are more suitable for volatile elements such as antimony, lead, mercury and tin. Here, the excitation is achieved by an rf field. These lamps, known as electrodeless discharge lamps, offer greater radiation intensities and have longer lives, but they are more bulky and require more complex power supplies.

The radiation from the lamp, of either kind, passes through a modulator before it enters the flame and hence interacts with the analyte. In its mechanical form, the modulator is simply a rotating disc with gaps around the

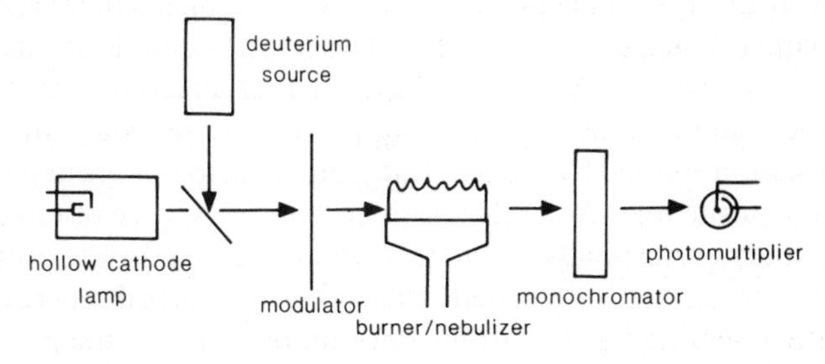

Figure 11.6 Basic components of atomic absorption spectrometer.

circumference, so that the beam of radiation is interrupted regularly, yielding pulses of radiation. This frequency of modulation can be incorporated into the detection system, so that detection readings can be taken with the lamp source effectively 'on and off'. These detector readings can then be compared and used to eliminate the radiation which arises from emission from the flame, so that only changes in the radiation from the source are determined. When the modulator shuts off the radiation from the source, any radiation reaching the detector must be due to emission, whereas when the beam is on, the detector will respond to emitted radiation plus that from the source, which will have been modified by passage through the flame. Alternatively, the beam may be modulated electronically by turning the power supply to the lamp on and off rapidly. This frequency of modulation is then similarly linked to the detection system, so that the 'radiation' and 'radiation off' detector responses can be determined. This information alone is not sufficient to calculate absorbance values for the metal atoms in the flame, as

$$\text{absorbance} = \log_{10}\frac{I_0}{I}$$

where I_0 is the intensity of the incident radiation *before* the sample and I is the resulting value after passing through it. Allowance for the I_0 value is usually achieved by setting the instrument to zero absorbance when water only, or a background solution not containing the analyte of interest, is being aspirated into the flame. This is exactly analogous to zeroing a single-beam UV–visible spectrophotometer (see Chapter 6). An alternative approach is to employ a double-beam arrangement where the reference beam passes through an air gap rather than the flame. However, in this case, in contrast to a double-beam UV–visible spectrophotometer, the reference and sample optical paths are not identical (only the sample beam passes through the flame) and so the instrument cannot be truly balanced. Nonetheless, variations in lamp output will affect both beams and so can be compensated for by this arrangement.

After passing through the flame, the beam of radiation passes through the monochromator, where all wavelengths other than that of the resonance line of the analyte are removed prior to measurement by the detection system. The monochromators used are often reflectance gratings affording a high degree of spectral resolution, which is variable (for example, bandpass 0·05–5 nm in UV). In many instruments, two complementary gratings are used to cover the entire UV and visible regions, as this provides a more uniform level of energy across the wavelength range. The photomultiplier tube is the most commonly used detector, as it provides adequate sensitivity over the required wavelength range.

The burners used in atomic absorption spectrometers are more commonly of the pre-mix type, and, in order to obtain good sensitivity, are long and narrow with the optical axis along the length of the burner. The most frequently used fuel/oxidant combination is acetylene/compressed air, although acetylene/nitrous oxide is also used where higher flame temperatures are required, as shown in Table 11.1.

11.3.1 *Interference effects*

Atomic-absorption spectrometry is subject to the same interference effects discussed for emission techniques. However, on account of the differences between what is actually being measured in the two techniques, the relative importance of the various interferences is somewhat different.

11.3.1.1 *Spectral interference.* The discrete nature of the radiation afforded by hollow cathode lamps and the high resolution (narrow bandpass) of the monochromator means that interference from other metallic species in the flame is very unlikely. In cases where this does occur, it is often possible to use a secondary emission line from the lamp, although this will lead to reduced sensitivity on account of the decrease in radiation intensity. The loss of sensitivity, however, may be acceptable in order to achieve the desired specificity.

The problem of background absorption from molecular species in the flame (and even scattering by particulate material) is less easily overcome. It is not

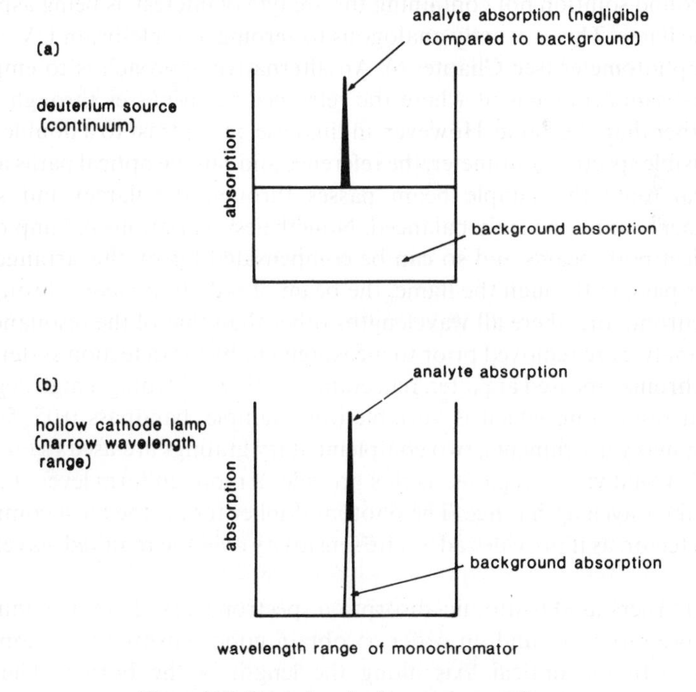

Figure 11.7 Deuterium background correction.

possible to scan the wavelengths around the absorption peak, as the radiation source only provides discrete wavelengths and so it is not possible to carry out a simple background correction as shown in Figure 11.4 for emission techniques. Automatic background correction is possible, however, by the addition of a further radiation source (Figure 11.6). A deuterium lamp is often used for this purpose, since it will provide a continuum (of wavelengths) either side of and including that of the analyte. Now as absorption by the analyte will only affect radiation of exactly the absorption wavelength, this means that the continuum from the deuterium lamp will only be *significantly* altered, as it passes through the flame, by background absorption, which occurs over a wider range of wavelengths (Figure 11.7*a*). This contrasts with the radiation from the hollow cathode lamp of discrete wavelength which will be directly affected by absorption in the flame of both the analyte and the background interfering species, (Fgure 11.7*b*). It must be remembered that the monochromator will allow a significant range of wavelengths to pass on either side of the discrete radiation from the hollow cathode lamp. By suitable switching of the two sources and comparison of the detector signals, it is therefore possible to remove the effects of background absorption continuously.

11.3.1.2 *Chemical interference.* The various processes which contribute to chemical interference (see Section 11.2.1) affect the number of free atoms of the analyte in the flame and therefore directly influence quantitative measurements in AAS. The methods by which these may be overcome, or at least reduced, apply equally well to absorption techniques as they do for emission. In general, higher flame temperatures are used in AAS than FES, but stable compound formation still remains a problem, as for example between calcium and phosphate ions. Ionization of excited atoms will also have a very small secondary effect on the population of ground-state atoms and this effect will be increased at higher temperatures. There may also be other ions formed in the flame which can suppress the ionization of the analyte and thus indirectly affect the ground-state population. The extent of ionization interference, which increases the ground-state population, is therefore temperature-dependent.

11.3.2 *Quantitative measurements*

The importance of AAS for analysis of metals in solution derives from its ability to produce precise and reliable quantitative data. However, this high degree of precision can only be turned into a similar level of accuracy when the various interference effects are taken into account, or eliminated. Thus, external calibration techniques should only be employed where interference effects are not present or the external standards have been prepared in a suitable medium, containing the interfering components in appropriate amounts. If this is not possible, for example if the sample matrix is not constant, then standard additions (Figure 11.5) will have to be carried out for *every* sample. This is a laborious, but essential, procedure which can be considerably facilitated by the use of automatic diluters and samplers.

11.3.3 *Alternative sampling techniques*

The inherent sensitivity of the atomic absorption process is to some extent lost in conventional flame techniques, as the ground-state atoms once formed are rapidly removed from the optical path, that is, they have a short residence time. Numerous attempts have been made to devise techniques which will increase the concentration of atoms in the light path, for a given amount of sample, and hence increase sensitivity.

The simplest methods use cups in which the sample solution is placed. The cup is brought near to the flame to evaporate the solvent and then finally placed in it for atomization. The analyte is released as atoms resulting in absorption over a short period, but of enhanced intensity. This technique has been refined further in the Delves cup, in which a tube is placed horizontally over the cup such that the atoms are forced to enter it through a hole and diffuse out along its axis, which is also the optical axis of the instrument. This procedure again increases the residence time of the atoms, and hence sensitivity, and has been widely used for the determination of lead in blood.

Mercury is unique in that it can be studied as a vapour at room temperature. It is generated from solutions of its ions by reduction, for example with sodium borohydride, and then swept by an air stream into an absorption cell placed in the optical path of the instrument. Any water vapour in the air stream is removed by a desiccant before the sample enters the cell. Certain other metals, such as arsenic, bismuth and tin, can form stable hydrides by a similar reduction process and these can then be swept into a conventional flame or heated quartz cell by an argon purge for absorption measurement (Godden and Thomerson 1980).

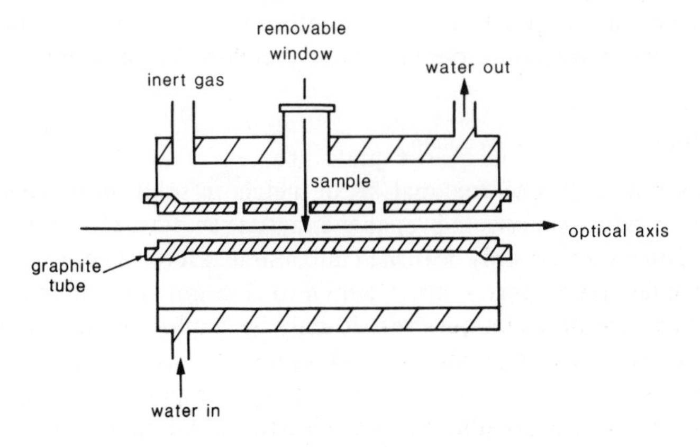

Figure 11.8 Graphite furnace.

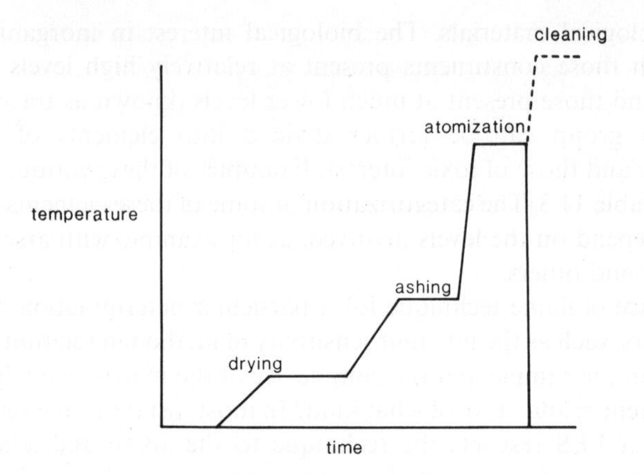

Figure 11.9 Typical temperature programme for graphite furnace.

The most important of the alternative sampling techniques is undoubtedly however the graphite furnace, where the flame is replaced by an electrically heated graphite tube, which lies along the optical axis of the instrument (Figure 11.8); the technique is thus often known as flameless AA. The sample, as a solution or solid, is introduced into the tube through a hole in the top surface. The tube is then heated to allow evaporation of the solvent, if necessary, ashing to destroy organic material and finally atomization. A typical temperature profile is shown in Figure 11.9. The exact temperatures and time intervals used for the various stages depend on the volatility of the element in question and also the nature of the organic matrix. Ideally, a high temperature should be used for the ashing stage as this will remove many of the components responsible for chemical interference or background absorbance. However, if the analyte of interest is relatively volatile, for example, lead, losses may occur at this ashing stage prior to atomization and therefore a compromise temperature must be used.

The increased residence time of the free atoms in the tube, before they are displaced by the inert gas passing through, again results in greatly increased sensitivity. For many elements, the detection limits are below 1 pg. The absorption signal produced by a graphite furnace is relatively short (1–2 s) and is different from that produced by a flame, which lasts as long as the sample is aspirated. It is therefore more precise to integrate the absorption over a fixed period in the former case, but this is unnecessary with a flame, as the continuous-intensity measurement can be recorded.

11.4 Applications

FES and AAS are vitally important techniques for the determination of a wide range of inorganic elements in biological samples, foodstuffs (Cowley, 1978) and physiological materials. The biological interest in inorganic elements covers both those constituents present at relatively high levels (known as minerals) and those present at much lower levels (known as trace elements). This latter group can be further divided into elements of nutritional significance and those of toxic interest. Examples of these various groups are shown in Table 11.3. The categorization of some of these elements is not clear and may depend on the levels involved, as for example with arsenic, cobalt, manganese and others.

The choice of flame technique for a particular determination depends on many factors, such as the inherent sensitivity of method in relation to the level of analyte in the sample and the complexity of the matrix, namely: are there any interferences and, if so, of what kind? In most instances, the relatively low sensitivity of FES restricts the technique to the alkali and alkaline earth elements, though fortunately these are more often present in biological samples at high levels (see Table 11.3). FES may also be used for the rare earth elements (such as lanthanum), where it has some advantages over AAS. The trace elements are more generally determined by AAS, where its inherent greater sensitivity can be used to full advantage. Routine surveys of toxic metals in foodstuffs or even plasma are carried out with a combination of flame and flameless AAS.

The nature of the sample preparation that is required prior to determinations by flame techniques depends on the nature of the matrix and the level of the analyte present in relation to those of potential interferents. Some of the more important preparation techniques are summarized in Table 11.4, together with some typical examples.

Table 11.3 Inorganic metallic elements of biological interest*

	Trace elements	
Minerals	*Nutritional interest*	*Toxic interest*
Calcium	Arsenic	Arsenic
Sodium	Cobalt	Beryllium
Potassium	Chromium	Cadmium
	Copper	Cobalt
	Iron	Chromium
	Iodine	Mercury
	Manganese	Manganese
	Molybdenum	Molybdenum
	Nickel	Nickel
	Selenium	Lead
	Silicon	Palladium
	Tin	Selenium
	Vanadium	Tin
	Zinc	Thallium
		Vanadium
		Zinc

*Data, in part, from Wolf (1982)

Table 11.4 Sample preparation techniques

Technique	*Basis*	*Example*
Dry ashing	Heat at 500–700 °C	Numerous applications, e.g. Fe in yeast
Wet ashing	Digestion with strong acids (H_2SO_4, HNO_3, etc.) at 250–300 °C	Numerous applications, especially where elements may be lost under dry-ashing conditions, e.g. Zn in saliva, semen
Precipitation	Removal of interferents by precipitation	Protein removal by precipitation with uranyl acetate in milk for Na determination
	Removal of analyte by precipitation	Removal of Ca by oxalate precipitation prior to its determination in flour digests
Complex formation	Chelation of metal ions to form complex extractable into organic solvents	Cu may be extracted as diethyldithiocarbonate complex into dichloromethane from ashed samples

The major limitation of FES and AAS, as usually employed, is that they are single-element techniques, and often multi-element determinations are required as, for example, in a survey of trace elements in soil samples. Multi-element analyses are possible with more elaborate techniques such as X-ray fluorescence or neutron-activation analysis. These are very powerful techniques relying respectively on the formation of excited atoms by short wavelength X-rays and a study of their characteristic emissions or the formation and subsequent characterization of radioactive nuclei formed by neutron bombardment. These are, however, beyond the scope of this text and the interested reader is referred to more detailed texts and monographs (Willard *et al.* 1981, Wolf and Harnley 1984).

Recent years have seen a resurgence in interest in emission techniques as a direct result of the introduction of inductively coupled plasma (ICP) sources. A plasma induced in an inert gas, such as argon, by rf energy can be maintained at a very high temperature. This results in the excitation of many elements and indeed emission from ions is often far more intense than from neutral atoms. ICP sources also provide the basis for multi-element techniques which have been used for many biological materials (Dahlquist and Knoll 1978, Jones *et al.* 1982).

References

Cowley, K.M. (1978) In *Developments in Food Analysis Techniques*, vol. 1, King, R.D. (ed.), Applied Science Publishers, London, 293.

Dahlquist, R.L. and Knoll, J.W. (1978) *Appl. Spectrosc.* **32**, 1.

Godden, R.G. and Thomerson, D.R. (1980) *Analyst* **105**, 1137.

Jones, J.W., Capor, S. and O'Haver, T.C. (1982) *Analyst* **107**, 353.

Willard, H.H., Merritt, L.L., Dean J.A. and Settle, F.A. (1981) *Instrumental Methods of Analysis*, 6th edn, Wadsworth, CA, 239–315.

Wolf, W.R. (1982) In *Clinical, Biochemical and Nutritional Aspects of Trace Elements*, Prasad, A. (ed), Alan R. Liss, New York, 427.

Wolf, W.R. and Harnley, J.M. (1984) In *Developments in Food Analysis Techniques*, vol. 3, King, R.D. (ed.), Elsevier Applied Science, London, 69–98.

Further reading

Cantle, J.E. (1982) *Atomic Absorption Spectrometry*, Elsevier Scientific, Amsterdam.

Christian, G.D. and Feldman, F.J. (1979) *Atomic Absorption Spectroscopy: Applications in Agriculture, Biology and Medicine*, Wiley, New York.

Dean, J.A. and Rains, T.C. (1975) *Flame Emission and Atomic Absorption Spectrometry*, vols. I–III, Marcel Dekker, New York.

Tsalev, D.L. and Zaprianov, Z.K. (1984) *Atomic Absorption Spectrometry in Occupational and Environmental Health Practice*, vols. I and II, CRC Press, Boca Raton.

Welz, B. (1985) *Atomic Absorption Spectrometry*, 2nd edn, VCH Verlagsgesellschaft, Weinheim.

12 Mass spectrometry

12.1 Introduction

Mass spectrometry is a powerful technique for the identification of pure compounds. It can also be used for confirmation of the purity of a sample and for quantitative analysis of mixtures. Ions are produced from sample molecules by various methods, including bombardment with a beam of electrons, and they are then accelerated and separated from ions of different molecular mass prior to detection. Often, the molecular or parent ion, which is produced from the sample molecule by loss of an electron, is observed together with a large number of ions (daughter ions) produced by characteristic fragmentations of the molecular ion. The molecular ion gives the relative molecular mass of the compound directly, while the fragmentation pattern provides a fingerprint which is used for identification of the molecule. The mass spectrum consists of a series of peaks of varying intensity plotted against the mass-to-charge ratio (m/z). Mass spectrometry combines high specificity with great sensitivity, since amounts of less than a picogram of some compounds can be detected.

12.2 The mass spectrometer

A mass spectrometer (Figure 12.1) contains an inlet system, which allows the introduction of a small amount of sample, usually in the gas phase; an ion source, allowing the production of ions in an evacuated chamber; slits across

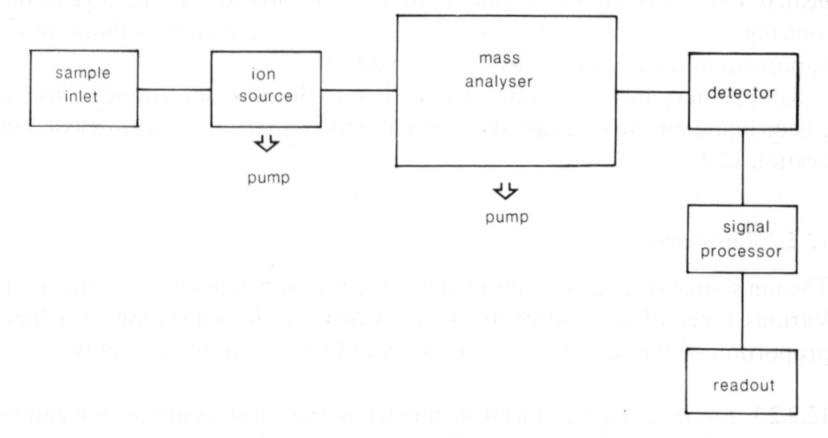

Figure 12.1 A block diagram of a mass spectrometer.

which a potential is applied to accelerate the ions; a mass analyser, which separates the ions on the basis of their mass-to-charge ratio; a detector; and a system for collecting and displaying data. The inlet system, ion source and mass analyser are evacuated to very low pressures (typically 10^{-6}–10^{-7} Torr) in order to avoid collisions of ions with other molecules which would cause neutralization or deflection and consequently spectra of reduced intensity and poor resolution. The vacuum is usually achieved by a combination of a rotary pump and an oil-diffusion pump, but recently turbomolecular pumps have replaced oil-diffusion pumps in some spectrometers.

12.2.1 *Inlet systems*

Gaseous samples can be admitted directly from a gas container through a valve into the evacuated ion source. Liquids are vaporized before being transferred to the ion source. In the case of moderately volatile liquids, the low pressure required for the ion source causes volatilization and the liquid may be introduced from an ampoule, which is cooled in liquid nitrogen during the evacuation of the spectrometer. On removal of the liquid nitrogen, the pressure is allowed to rise to the desired value and a valve is then closed to prevent introduction of excessive amounts of the sample. Less volatile samples may require the sample ampoule and inlet system to be heated. Heated inlet systems are often made of glass, since metal surfaces may catalyse decomposition at the high temperatures (up to 350 °C) that are sometimes necessary for vaporization. Other techniques for introducing liquid samples into a heated inlet system include injection with a microsyringe through a rubber septum, or the introduction of the sample in a capillary through a liquid metal (mercury or gallium) to glass seal.

Solids with relative molecular masses up to 1200 daltons may be introduced directly into the ion chamber on the end of a probe, which can be electrically heated. Extremely involatile samples may be introduced into the edge of the ionizing electron beam on the probe tip and ionization without prior vaporization may occur under these conditions.

Samples may also be introduced directly into the mass spectrometer from a gas or liquid chromatograph or a second mass spectrometer as discussed in Section 12.3.

12.2.2 *Ion sources*

The ion source of a mass spectrometer is the region where ions are generated. Various types of ion source are used to achieve the ionization of a high proportion of the sample molecules required for optimum sensitivity.

12.2.2.1 *Electron impact.* Electron impact is the most common ionization technique. The sample is vaporized and transferred through a slit into the ion

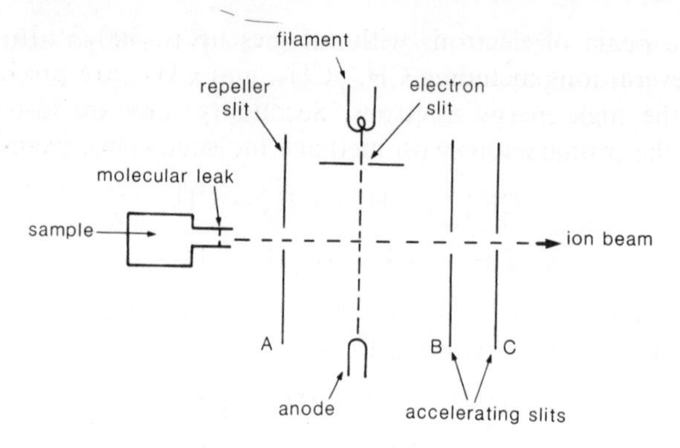

Figure 12.2 Electron impact ion source.

source which is at a pressure of about 10^{-6} Torr (Figure 12.2). The molecules are bombarded by a beam of electrons emitted from a heated filament, commonly tungsten or rhenium, and accelerated to a predetermined energy which may be varied up to 100 eV. The beam energy is dependent on the accelerating potential and an energy of about 70 eV is commonly used, since this is sufficient to cause fragmentation of the ions; this is often essential for structural identification. The ionization energies of most organic molecules are in the range 7–13 eV and beam energies within this range are used when it is desirable to minimize fragmentation of the ions in order to determine the relative molecular mass. The ionization energy is the energy required to remove an electron from a molecule to produce a charged radical called the molecular ion.

$$M + e^- \longrightarrow M^{\cdot +} + 2e^-$$

The molecular ion may fragment under high-energy electron bombardment by elimination of a radical (12.1), or by the loss of a molecule (12.2), in which all the electrons are paired:

$$M^{\cdot +} \longrightarrow A^+ + B^{\cdot} \qquad (12.1)$$

$$M^{\cdot +} \longrightarrow C^{\cdot +} + D \qquad (12.2)$$

An alternative process involving electron capture by a molecule is less probable than the loss of an electron, but it can occur under certain conditions. This leads to negative-ion mass spectrometry, which has some potential, but has not been extensively developed.

12.2.2.2 *Chemical ionization.* Chemical ionization involves the production of ions from a sample by chemical reaction of sample molecules with reagent ions formed from a reactant gas present in excess in a high-pressure ion source. The sample-to-reagent ratio is commonly 1:10^3–10^4. A reagent gas such as methane or ammonia at a pressure of about 1 Torr is ionized by bombardment with a beam of electrons with energies up to 300 eV. In the case of methane, several ions including CH_4^+, CH_3^+ and CH_2^+, are produced by the impact of the high-energy electrons. Secondary ions are also formed by reaction of the primary ions with methane molecules. For example:

$$CH_4^{\cdot+} + CH_4 \rightarrow CH_5^+ + CH_3^{\cdot}$$

$$CH_3^+ + CH_4 \rightarrow C_2H_5^+ + H_2$$

Collisions between sample molecules (RH) and reagent ions may often involve proton transfer (12.3) or hydride transfer (12.4):

$$CH_5^+ + RH \rightarrow RH_2^+ + CH_4 \quad (12.3)$$

$$C_2H_5^+ + RH \rightarrow R^+ + C_2H_6 \quad (12.4)$$

Chemical ionization produces a somewhat different spectrum if the reagent gas is changed. Since the reagent gas is present at a relatively high pressure in the ionization chamber, while a low pressure is required in the analyser, a high pumping capacity is required and a narrow slit at the entry to the analyser also aids the maintenance of the correct pressures.

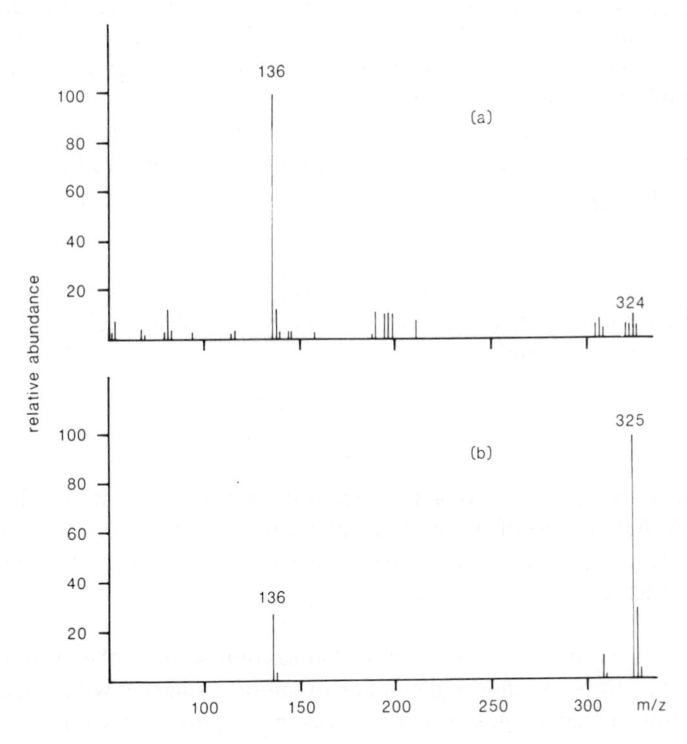

Figure 12.3 Mass spectra of quinine. (*a*) Electron impact; (*b*) chemical ionization. Redrawn with permission from Fales *et al.* (1970).

Chemical ionization has an advantage over electron impact in that the molecular ions (or pseudomolecular ions, for example, MH^+) formed have lower internal energies and therefore fragmentation is less extensive and detection of a molecular ion more probable. Thus, in the electron impact spectrum of quinine (Figure 12.3*a*) the molecular ion at m/z 324 is weak compared with the most intense peak of the spectrum (the base peak) at m/z 136. However, the chemical ionization mass spectrum (Figure 12.3*b*) shows a large pseudomolecular ion peak at m/z 325, with a relatively few weak peaks arising from ion fragmentation. The stability of the pseudomolecular ion MH^+ is also enhanced by its even-electron character, in contrast to the odd-electron molecular ion. The sensitivity of the technique is high.

12.2.2.3 *Field ionization and field desorption.* Field ionization is a procedure for the formation of ions by means of a powerful electric field. The sample molecules are passed between two closely spaced electrodes in an evacuated chamber (about 10^{-6} Torr) in the presence of an intense electric field (10^9–10^{10} V m^{-1}). Molecules that enter the electric field lose an electron by quantum-mechanical tunnelling. The anode consists of a point, blade or fine wire and an exit slit may serve as the cathode. Molecular ions are formed with little excess energy and fragmentation occurs only to a small extent. The total ion current, however, is relatively small (approximately 10 times less than by electron impact), due to the poor sample ionization efficiency and problems in focusing the beam of ions produced.

Field desorption is a modification of field ionization. It involves the ionization of molecules prior to vaporization. The sample is deposited in a thin film on the surface of the anode of a field ionization source, which can be achieved by dipping the anode into a solution after which the solvent is evaporated. The electric field is switched on and the anode is gradually heated with a small electric current. Under these conditions an electron may migrate from a sample molecule to the anode, which repels the resulting cation and

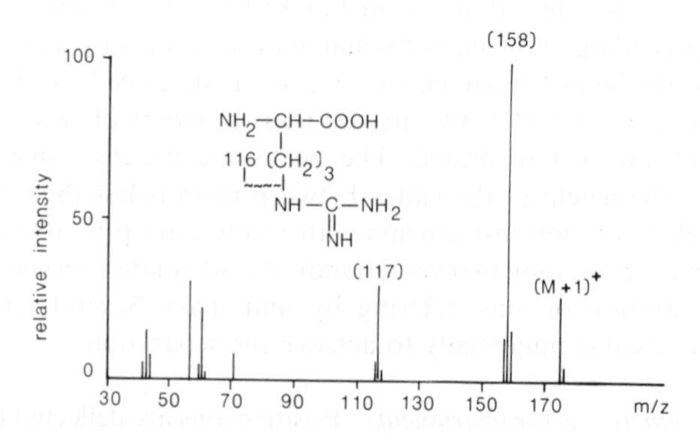

Figure 12.4 Field desorption mass spectrum of arginine. Redrawn with permission from Winkler and Beckey (1972).

causes the desorption of the ion. This technique is particularly suitable for biological molecules that decompose at the temperatures required for vaporization in other types of ion source. The enhanced intensity of the molecular ion (or pseudomolecular ion) peak compared with other ionization techniques is also valuable. For example, the electron impact and chemical ionization mass spectra of the amino acid arginine do not show a molecular ion peak, but a strong $(M + 1)^+$ peak is observed in the field desorption mass spectrum (Figure 12.4).

12.2.2.4 *Fast-atom bombardment.* Fast-atom bombardment is a recent development in sample ionization techniques which can provide mass spectra from samples of very high molecular mass. An inert gas (usually xenon) is ionized and the ions are accelerated to energies of 3000–10 000 eV. The charge on the ions is then neutralized by electron capture or charge exchange with low-energy xenon atoms, giving rise to accelerated inert gas atoms. The residual ions are removed by deflector plates and the accelerated atoms bombard the sample dispersed in a glycerol matrix. The bombardment gives rise to molecular ions, pseudomolecular ions and fragment ions. Fast-atom bombardment leads to ionization by 'sputtering' the sample into the vapour state. The phenomenon of sputtering is not fully understood, but it involves transfer of momentum from the accelerated atoms to molecules of the sample and glycerol matrix. This causes a series of collisions between molecules which allows some of them to gain sufficient energy to vaporize in the form of positively or negatively charged ions. Biologically active molecules, including polypeptides, with relative molecular masses up to 12 500 have been studied by this technique.

12.2.3 *Mass analysis*

The ions generated in the ion source are subjected to an accelerating potential applied across two slits (B and C in Figure 12.2). The beam of ions is then separated according to the mass-to-charge ratio of the ions. The ability of the analyser to distinguish between two masses is described by the resolution $M/\Delta M$ where M and $M + \Delta M$ are the mass numbers of two neighbouring peaks which are just separated. The peaks are usually considered to be separated if the height of the valley between them is less than 10% of their height. High-resolution instruments with resolution up to about 30 000 are available, but a resolution of several hundred is adequate for a spectrometer to achieve separation of ions differing by unit mass. Several types of mass analyser are used commercially to achieve the separation.

12.2.3.1 *Magnetic-sector instruments.* Positive ions are deflected by a magnetic field through an arc whose radius depends on the mass-to-charge ratio of the ions. The kinetic energy E imparted to ions of mass m and charge z by an accelerating voltage V is given by

$$E = \tfrac{1}{2}mv^2 = zeV \qquad (12.5)$$

where e is the electronic charge and v is the velocity of the ions. In a magnetic field of strength B, an ion following a path of radius r will experience a centripetal force of $Bzev$, which produces an acceleration of v^2/r. Now since force = mass × acceleration, we have:

$$Bzev = mv^2/r$$

Hence

$$r = \frac{mv}{Bze}$$

But (12.5) shows that

$$v = \left(\frac{2zeV}{m}\right)^{1/2}$$

and therefore

$$r = \frac{1}{B}\left(\frac{2mV}{ze}\right)^{1/2}$$

Hence the radius of the arc of deflection of an ion is dependent on the values of B and V. Ions of increasing m/z may be focused on the collector slit of a mass spectrometer either by varying the magnetic field B, known as magnetic scanning or, more usually, by electric or voltage scanning which involves varying the accelerating voltage V.

Early magnetic-sector instruments, including the first mass spectrometer which was reported by Dempster in 1918, deflected the ion beam through an angle of 180°, but recently 60° or 90° angles of deflection have become more

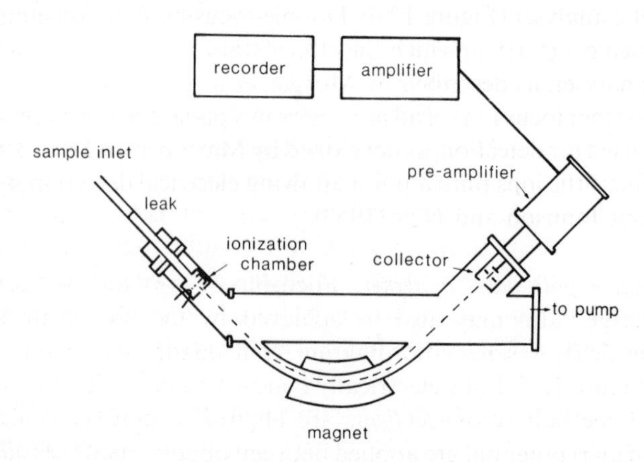

Figure 12.5 Magnetic sector mass spectrometer.

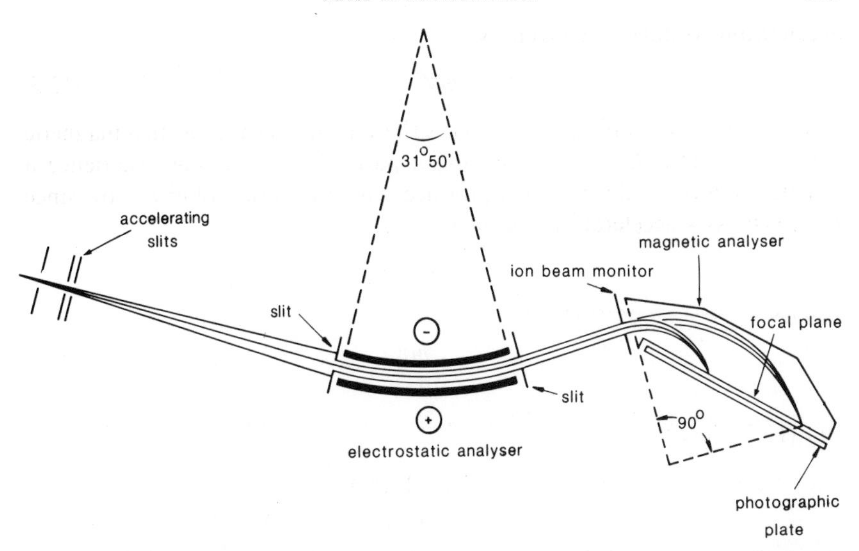

Figure 12.6 Double-focusing mass spectrometer with Mattauch–Herzog geometry.

common. Instruments of this type are cheaper to produce and the ion source and collector are well separated from the magnet region, allowing easier access for maintenance. A 90° magnetic-sector mass analyser is shown schematically in Figure 12.5.

Ions generated in the ion source vary somewhat in their kinetic energy and this produces a spread in the degree of their deflection in a magnetic field, which limits the resolution of the single-focusing magnetic-sector instrument described above. Resolution can be improved in a double-focusing spectrometer, which uses an electrostatic field to select ions with a narrow range of energy. The electrostatic field deflects the ions through a slit before they enter the magnetic analyser (Figure 12.6). Double-focusing spectrometers may also have reversed geometry in which the electrostatic analyser is located after the magnetic analyser, as described by Morgan *et al.* (1978). Commercial instruments may either focus ions of all m/z ratios in a plane allowing a photographic plate to be used for detection, as developed by Mattauch and Herzog (1934), or they may focus the ions onto a point allowing electrical detection by means of the design of Johnson and Nier (1953).

12.2.3.2 *Quadrupole mass analyser.* Focusing of ions on the basis of their mass-to-charge ratio may also be achieved by the use of an alternating quadrupole field. A schematic diagram of a quadrupole mass analyser is shown in Figure 12.7. Four electrically conducting ceramic rods (length 0·1–3 m) with hyperbolic cross-sections are aligned in parallel. A constant dc voltage and an rf potential are applied between opposite pairs of rods. Ions are transported along the x-axis of the quadrupole analyser and are subjected to a

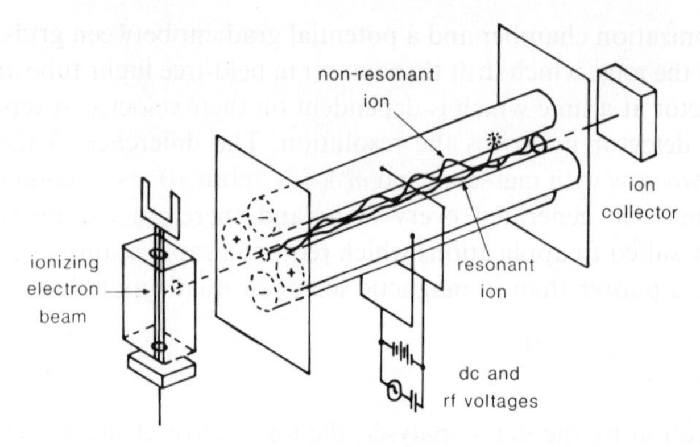

Figure 12.7 Quadrupole mass analyser. Redrawn with permission from Lichman (1964).

variable field causing a deflection in the *yz*-plane. Only ions with a specific m/z value have a trajectory which remains within the bounds of the electrodes for set values of the two voltages and rf. Ions of other m/z ratios are removed by impinging on the electrodes or escaping between the rods. The mass spectrum is scanned by varying the magnitudes of the two voltages keeping the ratio of their values constant, or by varying the frequency of the rf potential.

Quadrupole mass analysers are relatively inexpensive, compact and easy to operate and maintain. The absence of high voltages aids smooth operation. These analysers give good sensitivity up to m/z ratios of about 1000 and can scan rapidly. Mass selective detectors employing a quadrupole mass analyser are being widely used in gas chromatography, for which they are well suited. Magnetic analysers are more suitable than quadrupole analysers for high-resolution mass spectrometry and for the analysis of molecules with a high relative molecular mass.

12.2.3.3 *Time-of-flight mass analysers.* Time-of-flight mass analysis involves the separation of ions with uniform kinetic energy by virtue of their differing velocities (Figure 12.8). Ions are generated by electron impact using a short pulse of electrons, typically 1 μs. A voltage pulse on grid A withdraws the ions

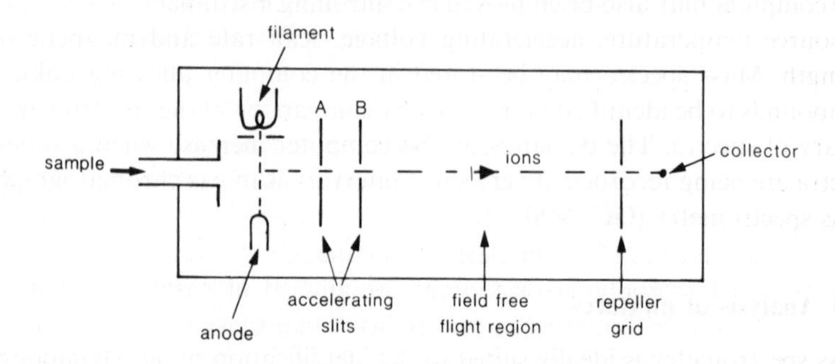

Figure 12.8 Time-of-flight mass spectrometer.

from the ionization chamber and a potential gradient between grids A and B accelerates the ions, which drift through a 1 m field-free flight tube and arrive at the detector at a time which is dependent on their velocity. A repelier grid before the detector improves the resolution. The difference in the time of arrival of two ions with masses m and $m + 1$ is about 10^{-7}s. A complete mass spectrum may be generated every 50 μs and therefore the instrument is particularly suited to applications which require a rapid scanning capability. Resolution is poorer than in magnetic sector or quadrupole instruments.

12.2.4 *Detectors*

After separation by the mass analyser, the ions arrive at the detector which may be either an electrical device or a photographic plate. Photographic plates coated with silver bromide in gelatin are capable of high sensitivity and are suitable for high-resolution instruments. Electrical devices include the Faraday cup and the electron multiplier. The Faraday cup measures the flow of electrons required to neutralize the ion current striking a small plate electrode, which is contained within a cup designed to prevent the escape of reflected ions and secondary electrons. The electron multiplier is the most common detector, since it is faster and more sensitive than the Faraday cup. The principle of detection is similar to that of a photomultiplier for a UV spectrophotometer (Figure 6.18). The ion beam from the mass analyser is accelerated by a potential difference and strikes the first plate of a multiplate ion multiplier. Each positive ion causes the emission of about two electrons from the plate and the electrons are accelerated and impinge on a series of plates or dynodes to achieve amplification of the current. Typical ion-multiplier tubes have about 20 dynodes prepared from a copper beryllium alloy. Current amplification of 10^6–10^7 can readily be achieved.

12.2.5 *Data handling and display*

Mass spectra may be displayed directly on a recorder, an oscilloscope or a galvanometer, but the use of a computer for data acquisition is most common. The computer may also be employed in controlling instrument variables, such as source temperature, accelerating voltage, scan rate and magnetic field strength. Mass spectra may be stored in the computer allowing unknown compounds to be identified more easily by comparison of the spectrum with a library of spectra. The demands on the computer increase when a series of spectra are being recorded at very short intervals as in gas chromatography–mass spectrometry (GC–MS).

12.3 Analysis of mixtures

Mass spectrometry is ideally suited to the identification of pure compounds. Quantification of simple mixtures can be obtained by determining the relative-mass spectral peak heights of one ion for each component and calibrating the response with a range of standard mixtures. However, quantification of complex mixtures requires a separation stage prior to mass spectrometry. Gas chromatography, high-performance liquid chromatography or a second mass spectrometer may be used as the separation stages. The components of very complex mixtures can be separated and identified by combining these techniques with a mass spectrometer. However, serious practical problems had to be overcome in the development of the combined instruments and it is only in recent years that commercial equipment has become available. As discussed in Section 3.8.6, the main problem in the development of GC–MS was the large difference in pressures between the chromatograph and the mass spectrometer. A 10^8-fold reduction in carrier-gas pressure must be achieved at the interface, with minimal loss of sample. This problem is more acute in liquid chromatography–mass spectrometry (LC–MS), where expansion from the liquid state to the gas phase occurs (about 200 times expansion at STP for chloroform rising to over 1000 for molecules such as water with a low relative molecular mass), in addition to the reduction of gas pressure. The maintenance of a low pressure in the source housing surrounding the ion source and in the mass analyser, and the transfer of solute to the vapour phase with minimum decomposition were major problems during the development of LC–MS.

Care must be taken in the design of the interface between the chromatograph and the mass spectrometer to minimize any band broadening due to adsorption effects or changes in the flow behaviour of the mobile phase.

The main requirement of the mass spectrometer is fast scanning capability, since the mass spectrum is scanned frequently during the chromatographic run. The system is connected to a computer which can store data arising from fast repetitive scanning of the chromatographic eluent. Sample detection involves the use of total ion-current or selected ion-current monitoring. The total ion chromatogram is the plot of total ion current against time and hence is analogous to the output from a flame-ionization detector. Components can be quantified by integration of the ion current by the computer.

12.3.1 *GC–MS interfaces*

Small-bore capillary columns can be coupled directly to the mass spectrometer, but wider bore capillary columns and packed columns are coupled via an interface which reduces the carrier-gas flow either by effluent splitting or by the use of a molecular separator. Effluent splitting can be achieved with a restriction such as a needle valve in the line leading to the mass spectrometer, with most of the sample being removed down a waste pipe. This procedure leads to poor sensitivity due to the analysis of only a fraction of the eluting components. The sensitivity can be improved by reducing the GC column outlet pressure, which enriches the sample in the carrier-gas stream.

Packed column GC–MS is best performed by enrichment of the eluent with a

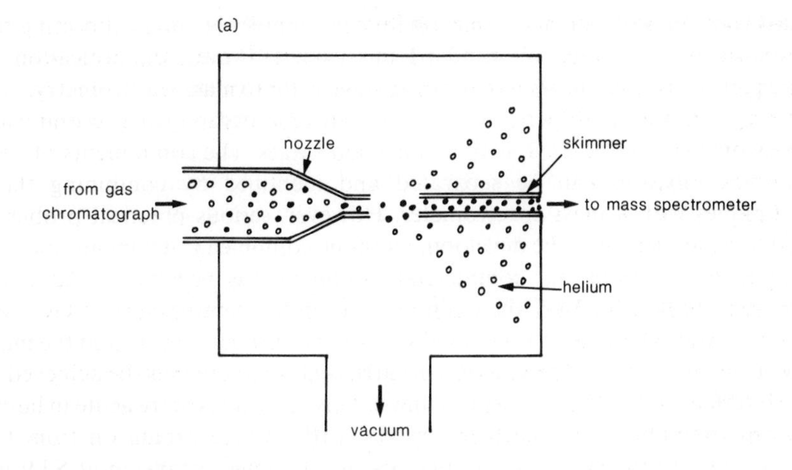

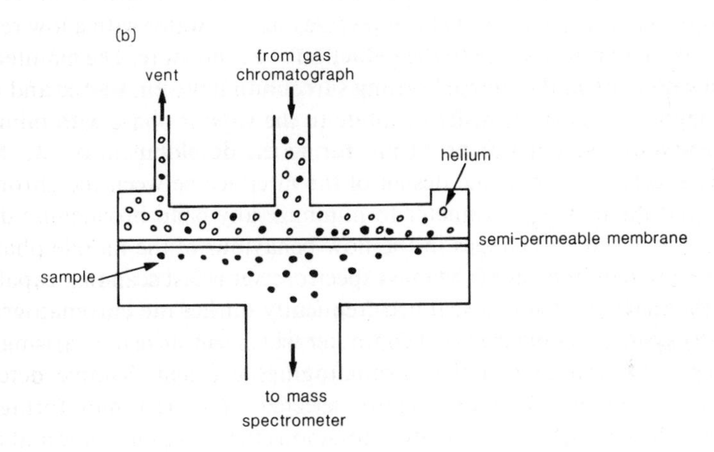

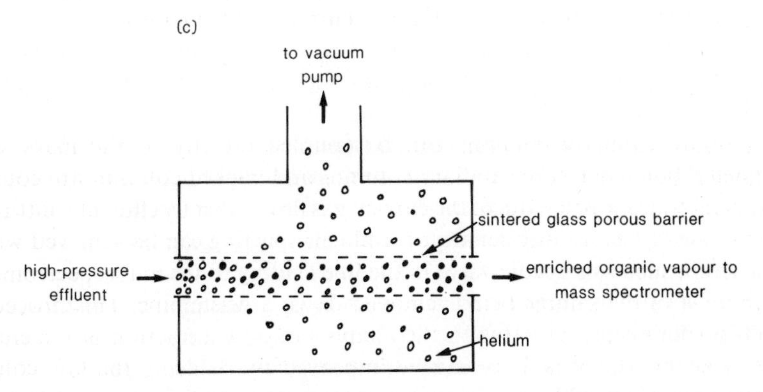

Figure 12.9 GC–MS interfaces. (*a*) Jet separator; (*b*) molecular separator using a semi-permeable membrane; (*c*) effusion separator.

molecular separator. This device selectively removes carrier-gas molecules from the gas flow entering the mass spectrometer. Three types of molecular separator are available: the jet separator, the semipermeable membrane and the effusion separator (Figure 12.9). The jet separator is very common on commercial instruments. It relies on differences in the rates of diffusion of gases in an expanding jet stream. The effluent from the gas chromatograph enters an evacuated chamber through a narrow orifice. The lighter carrier-gas molecules diffuse outwards more rapidly into the area of reduced pressure than the heavier organic molecules and they are pumped away preferentially. The heavier sample molecules move in the core of the jet stream towards the inlet of the mass spectrometer.

The semipermeable membrane separator is less common. It relies on the adsorption of organic sample molecules on a semipermeable silicone polymer membrane followed by diffusion of the organic molecules through the membrane to the mass spectrometer. The permeability of the membrane is determined by the product of the adsorption coefficient and the diffusion rate. Carrier-gas molecules such as helium may have a high diffusion rate, but they are not adsorbed by the membrane and therefore do not pass through it.

The effusion separator achieves enrichment by the use of porous glass or metal tubes through which the carrier-gas molecules diffuse preferentially.

12.3.2 *LC–MS interfaces*

The earliest LC–MS interfaces used either a moving-belt or direct-liquid introduction. In the moving-belt interface (Figure 12.10), the eluent from the chromatograph runs onto a continuous polyimide belt which is moving at a speed of 3–4 $cm\,s^{-1}$. The bulk of the solvent is removed as the belt passes under a radiant heater and the remaining solvent evaporates as the belt passes through two stages of vacuum which reduce the pressure to that of the ion source. The sample is evaporated from the belt by a radiant heater close to the site of ionization. The moving-belt interface is difficult to use with aqueous solvent systems and sample decomposition also limits its applications. Memory effects arising from incomplete evaporation of the sample may also be a problem.

In the direct-liquid introduction interface (Figure 12.10*b*), the chromatographic effluent is forced through a very small diaphragm orifice (2–5 μm diameter) which causes the formation of a stream of droplets. The droplets pass through a desolvation zone which is effectively an extension of the ion source. The electron beam ionizes the solvent vapour which then ionizes the solute molecules by chemical ionization. The flow of liquid entering the mass spectrometer is limited to about 20 $\mu l\,min^{-1}$, since the ion-source housing cannot be reduced to an acceptably low pressure at higher flow rates. A low flow rate can be achieved either by using columns of very low flow rate (microbore columns, 1 mm i.d.) or by splitting the flow from a conventional column. In the latter case, the sensitivity of detection is relatively low.

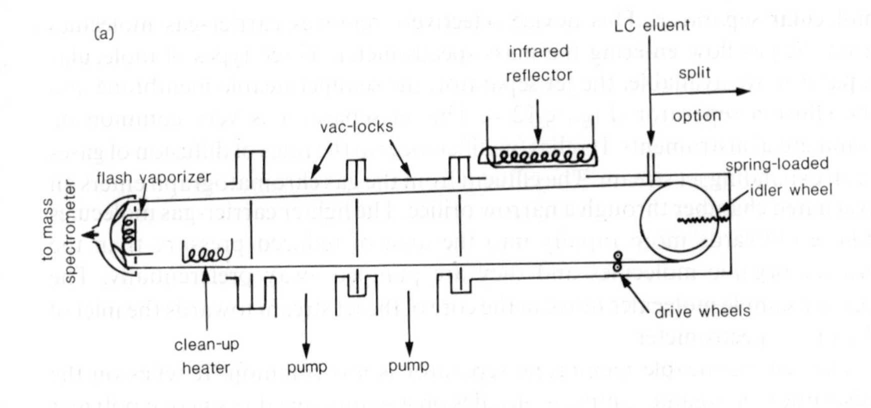

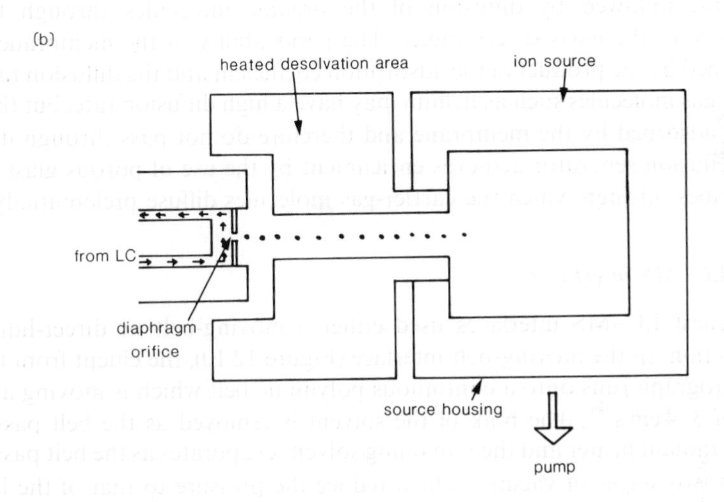

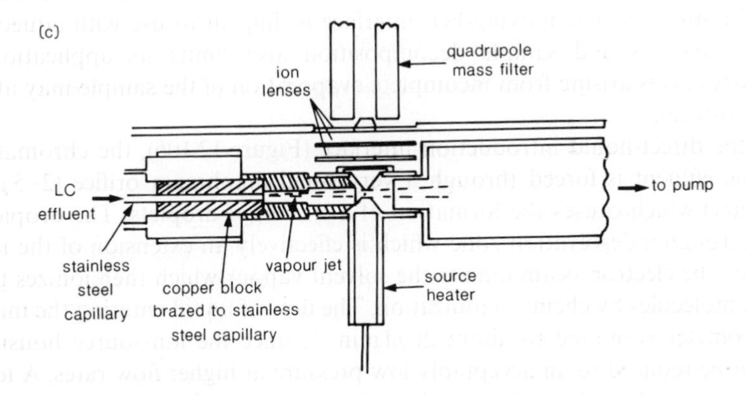

Figure 12.10 LC–MS interfaces. (*a*) Moving-belt (redrawn with permission from Millington, 1980); (*b*) direct-liquid introduction; (*c*) thermospray interface (redrawn with permission from Blakley and Vestal, 1983).

The thermospray interface (Figure 12.10*c*), which has recently been developed, allows the use of higher liquid flow rates than used in the moving-belt or direct liquid introduction interfaces. Most of the chromatographic solvent is vaporized as the solution passes through a heated narrow bore (150 μm) stainless-steel capillary. The vapour acts as a nebulizer gas and converts the remaining liquid into a jet of small droplets which emerges from the end of the capillary. The droplets then pass into the ionization chamber with evaporation reducing the diameter of the droplets prior to ionization. In the presence of an added electrolyte, such as ammonium acetate, the droplets possess a slight positive or negative charge. The field due to this charge is sufficient to cause the evaporation of sample ions from the very small droplets. Thus, thermospray ionization is a very mild process which allows the formation of molecular ions from polar compounds with a high molecular mass.

12.3.3 *Tandem mass spectrometry*

GC–MS and LC–MS rely on the separation of mixtures into individual components which can readily be identified by mass spectrometry. Tandem mass spectrometry achieves the same effect by linking two mass spectrometers together in order to achieve separation and identification of components. The first mass spectrometer ionizes all the components of the mixture by chemical or field ionization which generates mainly molecular or pseudomolecular ions. The molecular ions from the individual components of the mixture are separated by focusing successively on an exit slit according to their mass-to-charge ratio. Transmitted ions then pass into a chamber where fragmentation is induced by collision with gas molecules or by electron impact. The fragments are then separated and detected by the second mass spectrometer.

Tandem mass spectrometry is faster and more sensitive than a chromatographic process, but the cost of equipment is extremely high.

12.4 Determination of molecular structures

The mass spectrum gives the relative molecular mass of the compound directly if the molecular ion is observed. In the case of a high-resolution spectrometer, the molecular mass is given to within a thousandth of a mass unit and this allows the molecular formula of the compound to be determined unambiguously, since atomic masses do not have precise integral values. Thus caffeine, with a molecular formula $C_8H_{10}N_4O_2$ and mass 194·080, can be distinguished from a wide range of molecules including compounds with molecular formulae of $C_{10}H_{14}N_2O_2$ (mass 194·105), $C_{12}H_{18}O_2$ (mass 194·130) and $C_{13}H_{22}O$ (mass 194·167).

The molecular ion (M^+) represents the mass of the molecule containing the most common isotopes of the elements. However, significant amounts of other isotopes occur naturally (Table 12.1) and this gives rise to peaks at $(M+1)^+$

Table 12.1 Abundances of naturally occurring isotopes

Isotope	*% abundance*	*Isotope*	*% abundance*
^{1}H	99·985	^{16}O	99·76
^{2}H	0·015	^{17}O	0·037
^{12}C	98·892	^{18}O	0·204
^{13}C	1·108	^{33}S	0·76
^{14}N	99·63	^{34}S	4·22
^{15}N	0·37		

and $(M+2)^+$, which are smaller than the molecular ion. The heights of the $(M+1)^+$ and $(M+2)^+$ peaks relative to the molecular ion can allow the molecular formula to be determined. The relative height of the $(M+1)^+$ peak of a molecule $C_aH_bO_cN_d$ can be predicted from the natural abundances of the relevant isotopes to be $(1{\cdot}11a + 0{\cdot}015b + 0{\cdot}037c + 0{\cdot}37d)$. Thus, the $(M+1)^+$ peak of caffeine has a height of 10·41% of the molecular ion, while the alternative formulae $C_{10}H_{14}N_2O_2$, $C_{12}H_{18}O_2$, and $C_{13}H_{22}O$ have relative heights of 11·87%, 13·33% and 14·44% respectively. In the case of molecules containing chlorine, bromine, sulphur or silicon, characteristic peaks occur at m/z values of $M+2$, $M+4$, etc.

One useful observation in calculating molecular formulae from molecular ions is that molecules containing carbon, hydrogen and oxygen with zero or an even number of nitrogen atoms have even masses, while molecules with an odd number of nitrogen atoms have odd masses. This arises from the three-valent character of nitrogen.

Although the molecular ion, if observed, is useful in determining the relative molecular mass and molecular formula, full structural identification requires the analysis of the lower mass peaks in the spectrum which arise from fragmentation of the molecular ion. Three types of fragmentation process occur.

12.4.1 *Bond cleavage*

Cleavage of single bonds gives rise to ions arising from the loss of fragments of the molecule. Ions of mass $M - 15$ are commonly formed by the loss of a methyl group. Cleavage of a carbon–carbon bond is favoured by increased substitution of the carbon atoms, and the presence of a hetero atom also directs cleavage to the neighbouring carbon–carbon bond. In molecules containing carbon, hydrogen and oxygen only, the molecular ion has an even mass, while fragments formed by simple cleavage have odd masses.

12.4.2 *Rearrangements*

Loss of even mass fragments, such as water, can occur by rearrangement reactions. Thus, for alcohols, the molecular ion is commonly not observed, but an ion occurs at $M-18$ due to the loss of water:

$$\left[\begin{array}{l}R{-}CH_2{-}CH_2\\ \qquad\quad\;\; |\\ \qquad\quad\; OH\end{array}\right]^+ \rightarrow [R{-}CH{=}CH_2]^+ + H_2O$$

Often, rearrangement reactions have a high degree of structural specificity. For example, the McLafferty rearrangement leads to the loss of an alkene molecule from esters and ketones with a chain of at least three carbon atoms.

This rearrangement leads to a strong peak at $M-88$ for methyl esters of fatty acids, such as methyl stearate.

12.4.3 *Metastable peaks*

Bond cleavage and rearrangement reactions give rise to peaks in the mass spectrum at masses obtained by subtracting the neutral fragment from the molecular ion. However, this only applies if fragmentation occurs in the ionization chamber before acceleration of the ions. If fragmentation of the ion occurs in the field-free zone after the accelerating slits but before the mass analyser, the daughter ion is termed a metastable ion and occurs in the mass spectrum as a broad peak of low intensity at mass M_D, which is related to the molecular ion mass M, and the mass of the charged fragment M_F by the equation

$$M_D = (M_F)^2/M$$

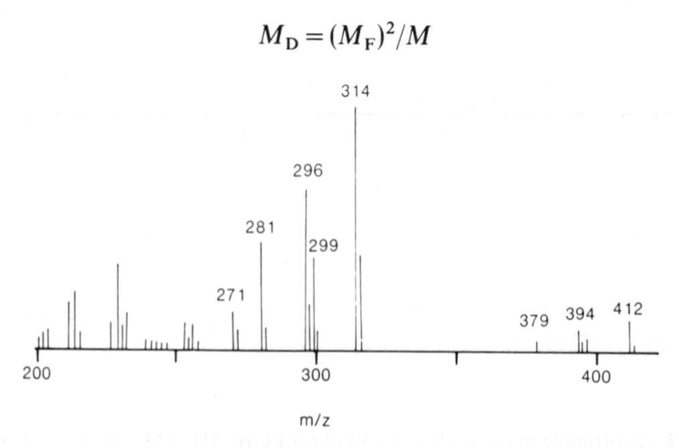

Figure 12.11 Mass spectrum of Δ^5-avenasterol. Redrawn with permission from Gibbons *et al.* (1968).

Metastable ions occur when the initial ion has a half-life of about 10^{-6} s, since this is sufficient for the ion to pass through the accelerating region but not through the mass analyser.

The occurrence of fragmentation ions can be illustrated by the mass spectrum of the phytosterol Δ^5-avenasterol (Figure 12.11). The molecular ion at *m/z* 412 fragments to give ions at *m/z* 397 by loss of a methyl group; 394 by loss of water; 379 by loss of a methyl group plus water; 314 by loss of part of the side chain C_7H_{14}; 299 by loss of C_7H_{14} plus a methyl group; 296 by loss of C_7H_{14} plus water; 281 by loss of C_7H_{14}, a methyl group and water; and 271 by the loss of the side chain and gain of 2 hydrogen atoms. The loss of C_7H_{14} is characteristic of sterols with a double bond at C24 due to the following rearrangement:

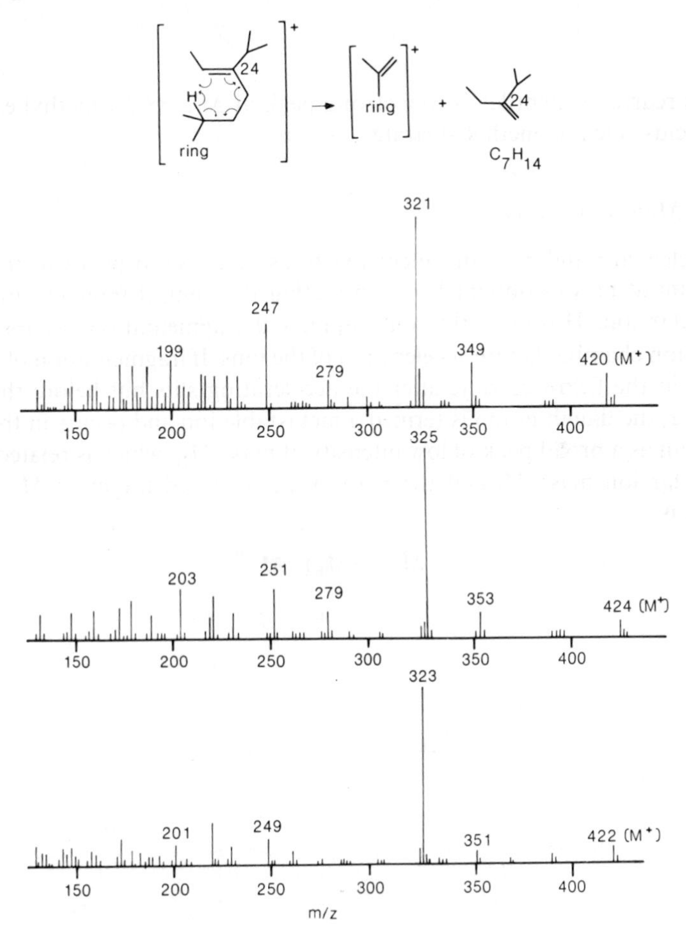

Figure 12.12 Quadrupole mass spectra (40 eV) of (*a*) PGB_2-ME-TMS, (*b*) 3, 3, 4, 4-$^2H_4$-PGB_2-ME-TMS, and (*c*) PGB_1-ME-TMS formed by derivatization of PGEs. Redrawn with permission from Ferretti and Flanagan (1979).

Mass spectrometry has made major contributions towards structure determination throughout the biological sciences. Identification of metabolites, peptide, protein and carbohydrate sequencing, structural elucidation of prostaglandins and analysis of food aromas are some of the areas in which the technique has had important applications. (Books describing areas of application in detail are given in 'Further reading'.) Molecules containing stable isotopes such as ^2H are often used in the application of mass spectrometry to the biological sciences. Isotopic substitution can reveal at which position in a complex molecule structural changes have occurred. Molecules containing stable isotopes can also be used as an internal standard in the quantitative determination of biological molecules. For example, Ferretti and Flanagan (1979) have employed (3, 3, 4, 4 – $^2H_4$) PGE_2 as an internal standard in the quantitative determination of the prostaglandins PGE_1 and PGE_2 by GC–MS. The prostaglandins were converted into the trimethylsilyl ether derivatives of the methyl esters of the PGB counterparts and quantified by selected ion monitoring (see Section 3.8.6) of the peaks at m/z 321 for PGE_2, m/z 323 for PGE_1 and m/z 325 for ($^2H_4$) PGE_2. The mass spectra of these compounds (Figure 12.12) show the justification for selecting these ions.

References

Blakley, C.R. and Vestal, M.L. (1983) *Analyt. Chem.* **55**, 750.
Dempster, A.J. (1918) *Phys. Rev.* **11**, 316.
Fales, H.M., Lloyd, A.M. and Milne, G.A. (1970) *J. Amer. Chem. Soc.* **92**, 1590.
Ferretti, A. and Flanagan, V.P. (1979) *Lipids* **14** (5), 483.
Gibbons, G.F., Goad, L.J. and Goodwin, T.W. (1968) *Phytochem.* **7**, 983.
Johnson, E.G. and Nier, A.O. (1953) *Phys. Rev.* **91**, 10.
Lichman, D. (1964) *Res. Dev.* **15** (2) 52.
Mattauch, J. and Herzog, R.F.K. (1934) *Z. Physik* **89**, 786.
Millington, D.S. (1980) *New Mass Spectral Techniques for Organic and Biochemical Analysis*, VG-Micromass Ltd, Altrincham.
Morgan, R.P., Beynon, J.H., Bateman, R.H. and Green, B.N. (1978) *Int. J. Mass Spec. Ion Phys.* **28**, 171.
Winkler, H.V. and Beckey, H.D. (1972) *Org. Mass Spectrum.* **6**, 655.

Further reading

Howe, I., Williams, D.H. and Bowen, R.D. (1981) *Mass Spectrometry: Principles and Applications*, 2nd edn, McGraw-Hill, New York.
Karasek, F.W., Hutzinger, O. and Safe, S. (1985) *Mass Spectrometry in Environmental Sciences*, Plenum, New York.
McLafferty, F.W. (1980) *Interpretation of Mass Spectra*, 3rd edn, University Science Books, Mill Valley, CA.
Merritt, C. Jr. and McEwen, C.N. (1979–80) Mass spectrometry, Parts A and B, in *Practical Spectroscopy Series*, vol. 3, Marcel Dekker, New York.
Rose M.E. and Johnstone, R.A.W. (1982) *Mass Spectrometry for Chemists and Biochemists*, Cambridge University Press, Cambridge.

13 Electrochemical techniques

13.1 Introduction

The electrical properties of a solid metal conductor, such as copper, can be defined in a simple manner by its conductivity, which is a measure of how easily electricity passes through it. The situation becomes more complex when the geometry of the conductor is changed, for example into a coil, and alternating current is used. However, under direct-current conditions the current passing through a conductor is defined by Ohm's law:

$$V = RI$$

where V (in volts) is the voltage across the conductor of resistance R (ohms) which results in a current of I (amperes). This simple treatment cannot be extended to the conduction of electricity, even as direct current, through solutions due to the presence of a number of additional effects. First of all, the nature of the solutions themselves, with respect to the presence of suitable conducting ions, and secondly the interaction between the electrodes placed in the solution and the solution components, must be considered.

13.1.1 *Nature of solutions*

Solutes may dissolve in water to form aqueous solutions of two distinct types. In cases where the solute molecules remain intact, molecular solutions are formed and any conductivity the solution may exhibit will arise only from the very limited ionization of the water itself:

$$H_2O \rightleftharpoons H^+ + OH^- \quad \text{(dissociation constant } K_w = 10^{-14}\text{)}$$

Compounds of this type would include organic molecules, which do not contain ionizable groups, such as acetone. Many other compounds, however, will ionize in solution, a process that takes place to a greater extent in more dilute solutions. This was shown mathematically by Ostwald who applied the law of mass action to the equilibrium:

$$BA \rightleftharpoons B^+ + A^-$$

so that

$$K = \frac{[B^+][A^-]}{[BA]}$$

or

$$K = \frac{\alpha c . \alpha c}{c(1-\alpha)}$$

$$= \frac{\alpha^2 c}{(1-\alpha)}$$

where c is the initial concentration of BA and $c(1-\alpha)$ is the remaining concentration after a fraction α has ionized. Clearly, if the equilibrium constant is to remain unchanged as c decreases (that is, greater dilution) then α must increase, and in the extreme case at infinite dilution, ionization would be complete. The ions now provide a means of conduction and values for α can be derived from conductivity measurements:

$$\alpha = \frac{\Lambda}{\Lambda_\infty}$$

where Λ is the molar conductivity (the conductivity corrected for concentration) of the solution in question and Λ_∞ is the molar conductivity for the same solute under conditions of infinite dilution (often obtained from the sum of ion conductivities, for example, $\Lambda_\infty NaCl = \Lambda_\infty Na^+ + \Lambda_\infty Cl^-$). The above predictions for the degree of ionization fit in well with those values determined for certain compounds, such as acetic acid, but not for others such as sodium chloride. The deviations for the latter group of compounds arise from the fact that they are already ionized within the solid-state crystal lattice, and solution simply involves a reduction in the attractive forces between oppositely charged ions, as a result of replacing the void between the ions with a medium of much higher dielectric constant. If these compounds are then completely ionized in solution, dilution should have no effect on the conductivity, once allowance has been made for the corresponding reduction in the total number of ions present (molar conductivity). However, in practice it is found that there is apparent increase in α with dilution. The nature of the ionic interferences, at high ion concentrations, which cause this effect is complex, but is associated with the fact that an ion of given charge will be surrounded by ions of opposite charge which will be moving in the opposite direction under the influence of an electric field. Additionally, water of hydration associated with these counter-moving ions will generate increased viscous drag, hence slowing the migration.

The conductivity of a solution is therefore related to the number of ions of solution and also their mobility, which is influenced by the presence of other species in solution as well as by the inherent properties of the ion in question, for example, charge, size and degree of hydration. The measurement of conductivity can therefore provide valuable information as to the 'ionic state' of the species present as well as their concentration.

13.1.2 *Electrode reactions*

When a metal electrode is placed in contact with a solution of its ions, one of two processes can take place. Metal atoms from the electrode can lose electrons and the resulting ions pass into solution,

$$M \rightarrow M^{n+} + ne^-$$

which results in an excess of electrons on the metal and hence a negative charge. Alternatively, the metal ions in solution abstract electrons from the metal and are deposited as metal atoms,

$$M^{n+} + ne^- \rightarrow M$$

a process which leads to a deficit of electrons on the metal and a net positive charge. In either of these cases a potential difference will become established between the metal electrode and the solution. The magnitude and sign of this potential will depend on the tendency for the metal ions to pass into solution. It should be noted that there are significant differences between this electrode process, resulting in an electrode potential, and ionization of an isolated atom, governed by the ionization potential. In the former case, the metal lattice must first be disrupted and then after ionization the ions will be hydrated as they pass into solution. The electrode potential is therefore a measure of the net effect of these three processes, lattice disruption, ionization and hydration.

In practice, it is impossible to measure the potential of an isolated electrode and so the difference in potential between two electrode systems is determined. The first problem that must be overcome is the establishment of a reference electrode to which all other electrode systems can be referred. The primary reference electrode which has been adopted is the hydrogen electrode (Figure 13.1), in which hydrogen gas at a pressure of 1 bar is bubbled slowly over a platinum electrode in contact with a hydrochloric acid solution of unit

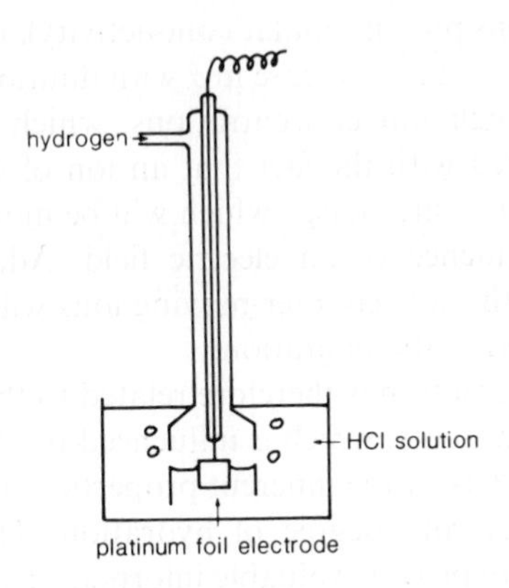

Figure 13.1 Hydrogen electrode.

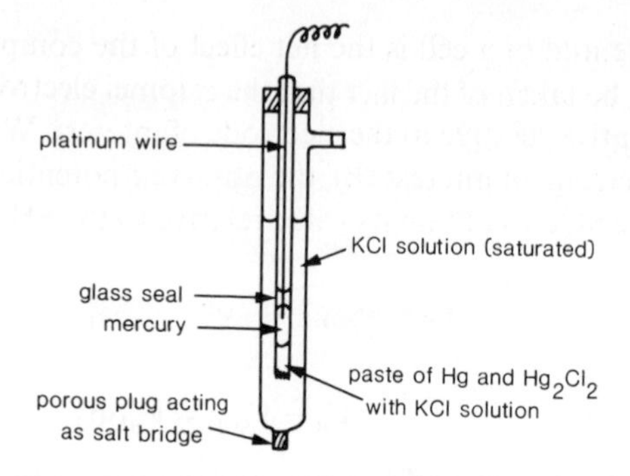

Figure 13.2 Calomel reference electrode (saturated).

H^+ activity (Ives and Janz 1961). This is a cumbersome unit to use and so a wide range of secondary reference electrodes (sometimes known as working or subsidiary reference electrodes) have been devised. Of the many types available, the saturated calomel electrode is probably the most widely used and is shown in Figure 13.2. The electrode reaction that takes place is $Hg_2Cl_2 + 2e^- \rightarrow 2Hg + 2Cl^-$ and so its potential is influenced by the concentration, or more precisely the activity, of the chloride ions. In addition to the saturated calomel electrode (saturated KCl solution), more precise electrodes are available with designated chloride concentrations (0·1 or 1·0 M) whose potentials are less temperature-dependent. Other reference electrodes, such as Ag/AgCl or Hg/Hg_2SO_4 are also encountered. Electrical contact between the reference electrode system and the electrode to be measured is usually completed by a salt bridge which is used to avoid introducing 'junction potentials' into the system. Such potentials arise from differences in mobility of ions across a phase boundary and may be considerably reduced by using an intermediate salt bridge containing, for example, potassium chloride solution whose cation and anion have similar mobility. The salt bridge is often incorporated into the reference electrode (Figure 13.2). The observed potential of the electrode pair (also known as a cell) must be corrected for the potential of the reference electrode, so that the potential of the electrode of interest, relative to that of the standard hydrogen electrode, can be established. This is a simple arithmetical procedure once the potential of the reference electrode (saturated calomel electrode, SCE, in this example) is known relative to the standard hydrogen electrode (SHE). This may be most easily illustrated diagrammatically:

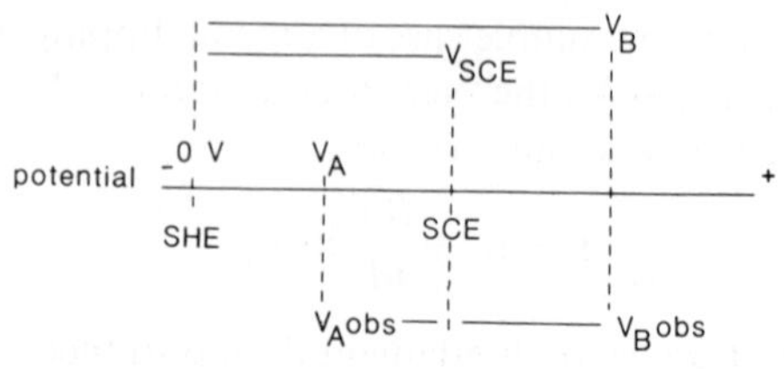

The observed potential of a cell is the net effect of the component electrodes and account must be taken of the fact that the calomel electrode potential may be positive or negative relative to the electrode of interest. When it is negative relative to the electrode of interest (B), the observed potential $V_B$obs is added to the value for the SCE to obtain its value relative to the SHE. With reference to the diagram

$$V_B - V_{SCE} = V_B\text{obs}$$

Thus

$$V_B = V_{SCE} + V_B\text{obs}$$

On the other hand, when the SCE is positive relative to the electrode under investigation (A), the observed potential $V_A$obs is subtracted from the value for the SCE

$$V_{SCE} - V_A = V_A\text{obs}$$

$$V_A = V_{SCE} - V_A\text{obs}$$

If electrode potentials are to be compared, for example as a measure of the tendency for reactions to take place, then they must be compared under standardized conditions. The above correction for the reference electrode potential is the first stage and this must then be followed by correction for concentration factors affecting the electrode process.

Considering a generalized electrode reaction:

$$\text{oxidized form} + ne^- \rightarrow \text{reduced form}$$

it has been established that

$$E = E^\circ - \frac{RT}{nF}\ln\frac{(\text{reduced form})}{(\text{oxidized form})}$$

or

$$E = E^\circ + \frac{RT}{nF}\ln\frac{(\text{oxidized form})}{(\text{reduced form})}$$

where E is the observed electrode potential (relative to SHE), E° is the standard electrode potential (that is, electrode potential under conditions of unit activity for oxidized and reduced forms), R is the gas constant, T is the absolute temperature and F is the Faraday constant. This is known as the Nernst equation and shows that the concentration (or strictly activities) of *all* the species involved in the electrode reaction will influence the observed potential. If we return to the simple case of a metal dipping in a solution of its ions, that is, $M^{n+} + ne^- \rightarrow M$, the Nernst equation can be simplified, as the activity of the solid metal is unity, so that

$$E = E^\circ + \frac{RT}{nF}\ln a_{M^{n+}}$$

The Nernst equation is of fundamental importance in understanding electrode reactions and its relevance will be seen again when particular techniques are studied.

When two electrode systems are combined to form a cell, the relative potential of the electrodes will be related to the tendency for the particular electrode reactions to take place. It is therefore not surprising to find that the electrode potential of a redox couple is related to the free-energy change and that the relative values of electrode potentials can be used to predict the course of redox reactions. This does not simply apply to redox reactions in which the electrode plays an active part, but also for redox couples in solution, for example, $Fe^{3+} + e^- \rightarrow Fe^{2+}$ in which case an *inert* electrode (such as platinum) placed in the solution will indicate its potential. A more positive reduction potential corresponds to a stronger oxidizing agent, since the compound or element has a greater tendency to be reduced (that is, a more negative change in free energy ΔG). For example, fluorine with a standard reduction potential of $+2{\cdot}87$ V ($F_2 + 2e^- \rightarrow 2F^-$) is a very strong oxidizing agent, whereas the alkali metals, for example sodium with a standard reduction potential of $-2{\cdot}76$ V ($Na^+ + e^- \rightarrow Na$) is a very strong reducing agent.

The cells described above, consisting of two coupled electrode systems, are often known as galvanic or voltaic cells as they produce a potential difference as a result of electrochemical action at the electrodes. A further type of cell must be considered in which an external voltage is applied to a pair of electrodes in solution. In this case, provided that the applied voltage is greater than that set up by the electrode processes, electrolysis will take place and the current passing through the system will lead to electrochemical reactions. The electrodes may play an active role in this process, as for example in the electrolytic transfer of copper from a copper anode (positive electrode) to a copper cathode (negative electrode) in a copper sulphate solution. In other cases, the electrodes may be inert, such as platinum, and electrolysis then involves reaction of the species in solution, as in the electrolysis of dilute acid solutions where oxygen and hydrogen are liberated from water. Such cells, in which electrochemical reactions take place as a result of an applied external potential, are known as electrolytic cells.

13.2 Conductivity of solutions

The conductivity of a solution of an electrolyte depends on both the number and type of ions in solution. This simple fact means that conductivity measurements have found a number of applications as the basis of analytical methods. First of all, the measurement of conductivity must be briefly considered.

13.2.1 *Measurement of conductivity*

The solution of interest is placed in a conductivity cell such that a fixed volume of the solution resides between the electrodes, that is, with fixed electrodes. In

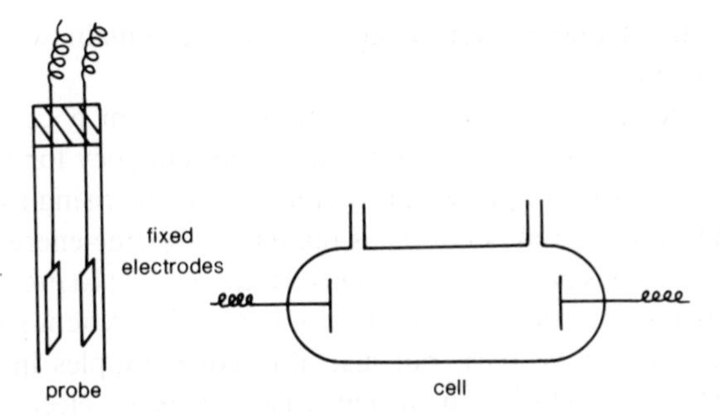

Figure 13.3 Conductivity cells.

addition to cells, it is also possible to use a conductivity probe which also contains fixed electrodes (Figure 13.3). The resistance of the cell is then measured ($R\Omega$) with an ac potential (usually about 1000 Hz), to avoid complications due to electrolysis at the electrodes. The effects of electrolysis are further reduced by using platinized platinum electrodes (platinum electrodes with an electrolytically coated layer of finely divided platinum (black)). Measurement of resistance was traditionally carried out with a Wheatstone bridge with earphones to detect the balance point, but improved electronics now means that a direct-reading instrument can be used. In order to convert the observed resistance ($R\Omega$) into conductivity (κ), the physical dimensions of the cell must be known, in particular the cross-sectional area of the electrodes (a) and the distance between them (l), then

$$\kappa = \frac{l}{Ra}$$

Now in practice these dimensions may not be very easy to measure and so it is more usual to determine the ratio l/a (cell constant) by measuring another solution of known conductivity. This is most commonly potassium chloride as it can be easily obtained in a high degree of purity, so if κ_{KCl} and R_{KCl} are the respective values for a potassium chloride solution of known concentration, then

$$\kappa_{KCl} = \frac{l}{R_{KCl}a}; \qquad \frac{l}{a} = \kappa_{KCl}.R_{KCl}$$

and

$$\kappa = \frac{\kappa_{KCl}.R_{KCl}}{R}$$

It must be noted that this simple procedure will only work if the cell dimensions are constant, thus the electrodes must not be moved or damaged in any way.

The conductivity value obtained may then be converted into a molar (or equivalent) conductivity to remove the effects of dilution.

$$\text{Molar conductivity } \Lambda = \kappa / c$$

Great attention must be paid to the units employed and a wide range will be encountered: $\Omega^{-1}\,m^2\,mol^{-1}$ (where c is given in $mol\,m^{-3}$ and κ in $\Omega^{-1}\,m^{-1}$) or $\Omega^{-1}\,cm^2\,mol^{-1}$ (where c is given in $mol\,cm^{-3}$ and κ in $\Omega^{-1}\,cm^{-1}$).

13.2.2 *Analytical applications*

The determination of conductivity values may be used to calculate a wide range of physical parameters of electrolytes, including hydrolysis constants, dissociation constants and solubility products, and the interested reader is referred to standard texts (for example, Ramette 1981).

Here we shall discuss some of the more analytical applications under the headings of direct methods and conductometric titrations.

13.2.2.1 *Direct methods.* The conductivity of a solution is due to the combined effects of all the ions present and so conductivity cannot, on its own, provide any qualitative data as to the nature of the ions present or indeed the concentrations of particular ions. However, in certain instances where the nature of the ions is known or is not of interest, conductivity may be used to estimate their concentration. A simple example of this would be in monitoring the deionization of water to produce pure water for analytical purposes. The nature of the ions initially present would be known approximately and the fall in conductivity during the process can be used to follow its progress. In this example it could also be argued that knowledge of the nature of the ions is not essential as the presence of *any* ions would increase the conductivity and this is undesirable. Another similar example would be the detection of electrolytes in products where they should be absent. Glucose syrups are prepared from starch by acid or enzymic hydrolysis; after the former, the excess acid is neutralized and the salts formed removed by ion exchange. If this process is successful, an aqueous solution of the syrup should have only a very low conductivity (glucose does not ionize in solution) and any abnormally high value would show immediately that salts were still present.

13.2.2.2 *Conductometric titrations.* The use of conductivity measurement for the determination of end points in volumetric analysis is one of its more important applications, especially in those instances where coloured solutions preclude the use of indicators. In all titrations, there is a change in the type of ions present in solution and often also a change in the number of ions, but conductivity measurement will only be possible for end-point determination when there is a significant change in conductivity over the end-point region. The shape of the plot of conductivity against volume of titrant will depend on changes in ions in solution and we can consider a number of simple cases.

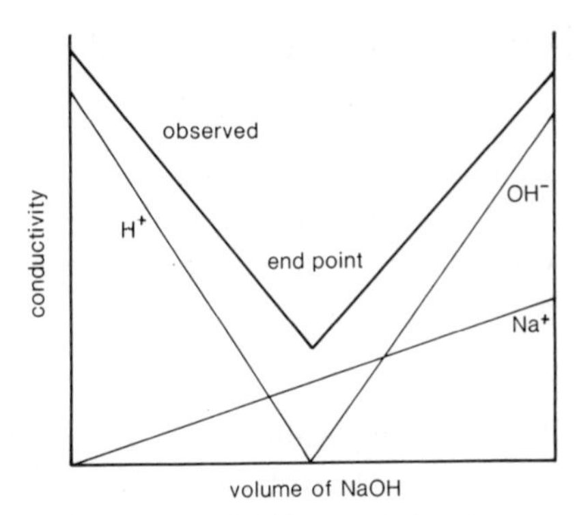

Figure 13.4 Conductometric titration of hydrochloric acid with sodium hydroxide (strong acid–strong base). Observed conductivity and contributions from individual ionic species.

13.2.2.3 *Strong acid–strong base.* This is the simplest case and probably the most amenable to conductometric titration. Consider, for example, the titration of hydrochloric acid with sodium hydroxide. At the start, the solution will contain H^+ and Cl^- ions, and, as the sodium hydroxide is added, the added OH^- ions will combine with H^+ to form essentially non-conducting H_2O, and the Na^+ ions will remain in solution with Cl^-. Thus, effectively the only change apart from dilution is the substitution of Na^+ ions for H^+. The H^+ ions have a much greater equivalent conductivity than Na^+ ions, due to their greater mobility, and therefore the conductivity will drop until the end point is reached. After complete neutralization of the acid, the conductivity will rise again due to the presence of excess Na^+ and OH^-. If we ignore dilution effects, a simple graph of two intersecting straight lines would be formed, indicating precisely the end point (Figure 13.4). The contribution of the individual ions is also shown in the same diagram, again ignoring effects of dilution.

13.2.2.4 *Weak acid–strong base.* The situation with a weak acid is more complex and the exact shape of the resulting titration curve depends on the dissociation constant of the acid as well as its concentration. In the case of acetic acid (K_a 1.8×10^{-5} or pK_a 4.8), at the start of the titration only a small fraction of the acid will be dissociated

$$CH_3CO_2H \rightleftharpoons CH_3CO_2^- + H^+$$

hence the conductivity will be low. As sodium hydroxide is added, the H^+ ions will be initially removed to form water and then more acetic acid will dissociate; however the concentration of acetate ions will have been increased and so the ionization will be suppressed. This results in a decrease in the number of H^+ ions in solution and therefore a decrease in conductivity. As the titration proceeds, the increased concentrations of Na^+ and acetate ions

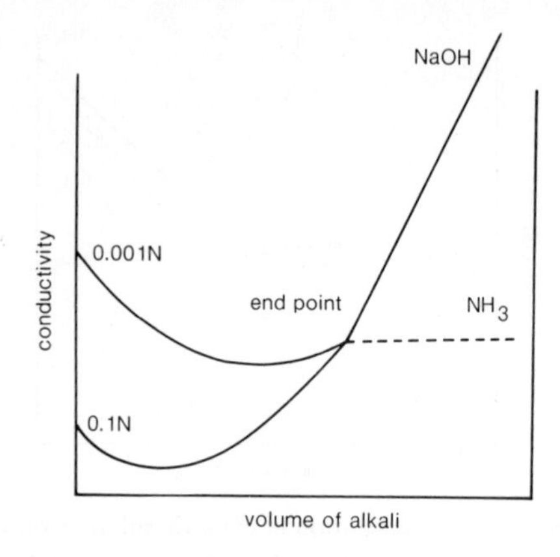

Figure 13.5 Conductometric titration of acetic acid with sodium hydroxide or aqueous ammonia.

override this effect and so the conductivity increases, until the end point is reached when the excess Na^+ and OH^- ions will produce an even greater increase. The exact shape of the earlier portion of the curve, and to some extent the precision of the end point, depends on the concentrations employed. However, in all cases it is clear that the end point is very poorly defined (Figure 13.5). The problem can be partially solved by using a weak base instead, such as ammonium hydroxide (aqueous ammonia) which, as it is only slightly dissociated, does not cause an increase in conductivity after the end point (Figure 13.5). It is interesting to note that whereas a weak base helps in the end-point detection with conductivity, it is less satisfactory than a strong base when pH is being monitored.

13.2.2.5 *Mixed acids.* A further interesting example is the titration of mixed acids, that is, a mixture of a strong and a weak acid, as for example a sample of

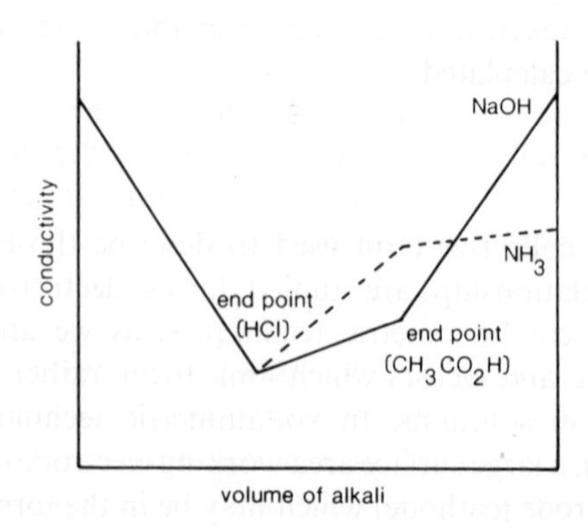

Figure 13.6 Conductometric titration of mixed acids (HCl and CH_3CO_2H) with sodium hydroxide or aqueous ammonia.

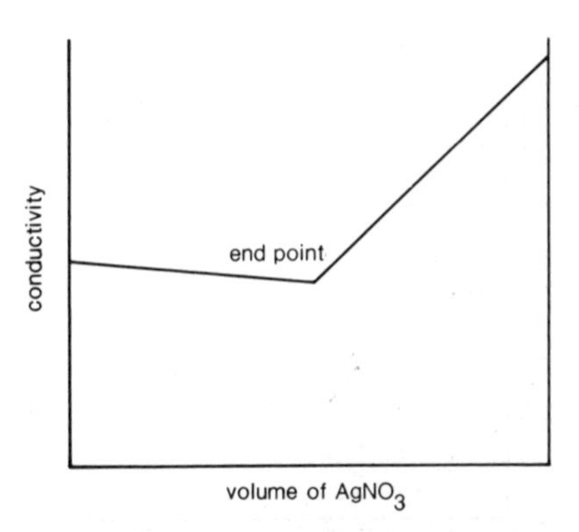

Figure 13.7 Conductometric titration of chloride solution with silver nitrate.

vinegar (acetic acid) adulterated with a strong acid (sulphuric acid). The resulting conductivity curve is shown in Figure 13.6 and the shape should be readily understandable as it is a composite of Figures 13.4 and 13.5. Here again, in certain cases, a more precise end point could be realized by using a weak base.

13.2.2.6 *Precipitation reactions.* A further group of reactions which is often encountered in volumetric analysis involves the removal of ions from the system by precipitation; of these, those involving silver nitrate reacting with halides are the most important. Consider the titration of a salt (NaCl) solution with $AgNO_3$. Initially only Na^+ and Cl^- ions are present; as the titration proceeds the chloride ions are removed by precipitation of AgCl, and NO_3^- ions remain in solution with Na^+. Now as the equivalent conductivities of Cl^- and NO_3^- ions are very similar, only a slight change in conductivity will be noted up until the end point, after which there will be an increase due to the presence of excess Ag^+ and NO_3^- ions (Figure 13.7). Changes in total solution volume should be taken into account to enable more accurate changes in conductivity to be calculated.

13.3 Voltammetry

Voltammetry is a collective term used to describe those techniques where voltage–current relationships are studied during electrolysis. These must not be confused with conductometric techniques, as we are now considering electrode reactions, and factors which limit them, rather than bulk conduction due to ions in solutions. In voltammetric techniques, two different electrodes are used; a large surface area working electrode (anode) and a much smaller microelectrode (cathode) which may be in the form of small drops of mercury (dropping-mercury electrode) or a rotating platinum electrode. As a current passes through these two electrodes, significant differences will be

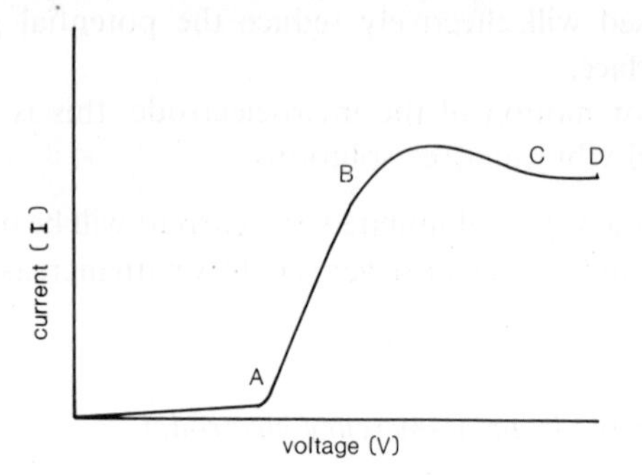

Figure 13.8 Current/voltage curve under static conditions at a microelectrode (cathode) showing decomposition potential (*A*) and diffusion-controlled current ($C \rightarrow D$).

observed in the immediate vicinity of the electrode surfaces. The working electrode has a large surface area and therefore the current density will be very small with the result that the concentration of ions in the layer of solution over the electrode will remain constant, which in turn will give a constant potential. On the other hand, at the microelectrode, because of its much smaller surface area, there will be a much higher current density and therefore a rapid depletion of ions in the adjacent solution and the current passing through the cell will fall. The electrode is then described as polarized.

Consider the example of the two electrodes placed in a very dilute (0.01 M) Cu^{II} solution. As the voltage applied to the system is increased (Figure 13.8), the current initially stays low until a point is reached (A) where the decomposition potential of Cu^{II} ions in 0·01 M solution has been reached, and the ions are electrolytically deposited on the microelectrode (cathode). It should be noted that under these conditions, the decomposition voltage, corresponding to point (A), is not constant for a given element, but will vary with concentration as dictated by the Nernst equation; the decomposition potential is effectively that which is required to overcome the potential formed by the Cu/Cu^{II} electrode system. As the voltage is further increased, the current increases along AB. Now if the solution was adequately stirred and there was therefore no effective depletion of ions near the electrode, the current would continue to rise. However, in the static situation, the concentration of ions near the electrode will drop and so therefore will the current to some limiting value (Figure 13.8). The current will then depend on:

(i) Diffusion of ions as a result of the concentration gradient formed by depletion governed by Fick's law, which states that the rate of diffusion is proportional to the concentration gradient. This is the basis of using voltammetry as a quantitative technique.
(ii) Migration of charged ions in the electric field. This is often reduced by adding an additional electrolyte such as potassium chloride. The K^+ ions will crowd around the cathode and, as they cannot be discharged with the

potentials used will effectively reduce the potential gradient near the electrode surface.

(iii) Convection or motion of the microelectrode, this is reduced by using unstirred and vibration-free solutions.

Thus, if the (i) and (ii) are eliminated, the current will be limited by diffusion at some point (C) and will only rise very slightly with increased potential along (C → D).

13.3.1 *Polarography (dropping-mercury electrode)*

Polarography, with a dropping-mercury electrode, is still the most widely used voltammetric technique. The basic apparatus required is shown in Figure 13.9. Mercury passes through a fine glass capillary to produce small drops every 3–5 s. The working electrode consists of the pool of mercury residing in the base of the cell, which, if chloride ions are present in the electrolyte, will act as a calomel electrode. The associated electronics must simply be able to provide a potential difference of between $+0{\cdot}4$ V to $-1{\cdot}8$ V versus the calomel electrode. Above $+0.4$ V, mercury dissolves and below $-1{\cdot}8$ V, gas evolution takes place in acid solution and other electrolytes are discharged. The dropping-mercury electrode has considerable advantages over other microelectrodes, in particular:

(i) its surface area is reproducible and can be easily determined by weighing a number of drops

(ii) its surface is constantly renewed and so no problems of surface poisoning arise. The mercury may be re-used after acid washing

(iii) the diffusion current rapidly reaches a steady value

(iv) mercury forms amalgams with many metals which causes a decrease in their reduction potentials

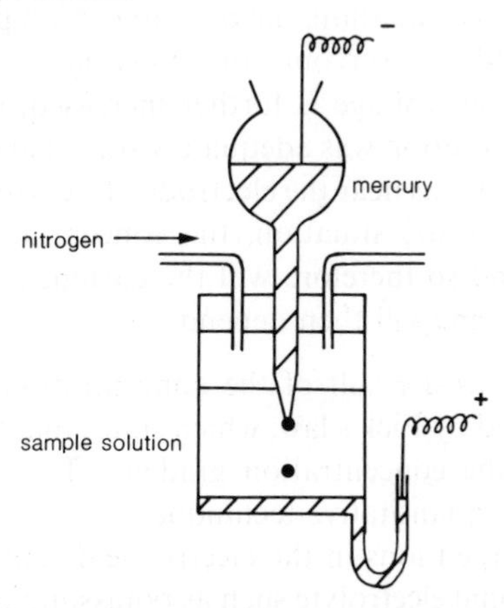

Figure 13.9 Dropping mercury electrode.

(v) hydrogen has a high overpotential on mercury which means that other elements, whose reduction potential is lower (more negative) than that for the reversible hydrogen discharge, can still be reduced.

As the drop of mercury forms on the capillary tip, the current passing through the cell will increase under a given potential to reach a maximum just before the drop falls. The current will then effectively fall to zero before the next drop forms. If a recording device were to be used with a very short time constant (that is, very rapid response) a trace with wild oscillations would be produced and in practice it is usual to employ a considerable degree of electronic dampening. A typical polarogram is shown in Figure 13.10 with the important parameters characterizing the trace marked. It should be noted that although the decomposition potential normally depends on the concentration of reducible ions in solution, under conditions of diffusion-controlled migration, it can be shown that the half-wave potential is independent of this concentration and the value is characteristic of the reacting material. Furthermore, the wave height (diffusion current) is related to the concentration of reacting material and this value is used for quantitative measurements.

There are three commonly used methods of quantification based on wave-height measurements. In the first, a calibration curve is produced with standards covering the range of concentrations expected in the samples. The calibration is often linear, but for accurate work the standards used should include concentrations both above and below the sample values. The second method involves the introduction of an internal standard to the sample, with subsequent comparison of wave-height values. This is not widely used, owing

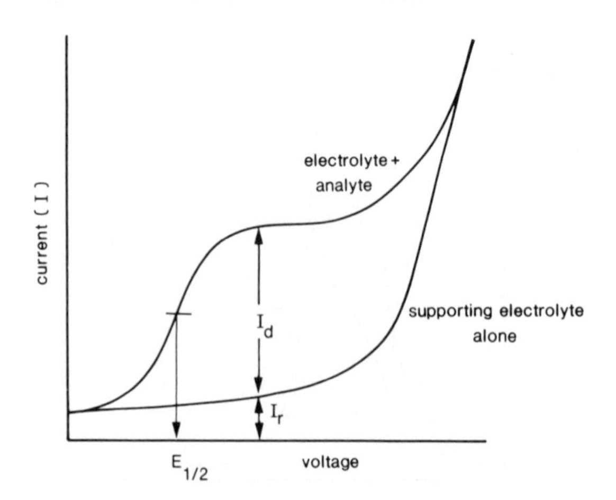

Figure 13.10 Idealized polarogram obtained with dropping mercury electrode (pulses not shown). $E_{1/2}$ is half-wave potential, characteristic of analyte; I_d is diffusion current of analyte; I_r is residual current from electrolyte.

to the requirement that the internal standard must have a significantly different half-wave potential (by about ± 0·2 V). The final method makes use of standard additions where the solution of interest is measured and then remeasured after addition of a known amount of the substance of interest. The difference in wave heights is then proportional to the amount added, from which the concentration of the original solution can be determined. This is probably the method of choice as the presence of any interfering compounds will affect the standard in the same way and so will have less effect on the final result.

In solutions which contain two or more reducible substances with well-separated half-wave potentials (that is, differences greater than 0·15 V), distinct waves are formed corresponding to each component, which may then be precisely determined. However, in cases where ions have rather similar potentials and distinct waves are not formed, quantitative data may still be obtainable using derivative techniques, where the slope of the wave is monitored. The example shown in Figure 13.11 is a limiting case where even the derivative curve is not completely resolved, but clearly any quantitative measurements based on this will be superior to those from the original wave, where the presence of more than one substance can only just be seen.

In addition to polarography carried out under direct-current conditions, there also exist several modified techniques such as pulsed polarography, oscillographic polarography and chronopotentiometry (measurement of the time necessary for the voltage to fall under constant current). These are specialized techniques beyond the scope of this book, but details may be found in appropriate texts (Bassett *et al.* 1983).

However, of these techniques, anodic stripping voltammetry is worthy of mention as it is widely used in the determination of heavy metals (cadmium,

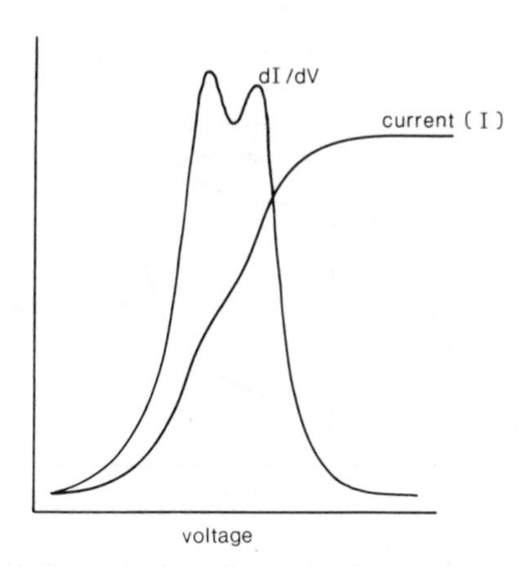

Figure 13.11 Polarogram showing the improved resolution of two analytes with similar $E_{1/2}$ values afforded by derivative techniques (pulses not shown).

lead) in biological and food samples (Murthy 1974). In this technique, a potential (0·2–0·4 V more negative than the highest reduction potential of components to be encountered) is applied to a mercury-drop electrode (a single drop of mercury stationary at the capillary tip) so that a significant proportion of the ions in solution will be discharged on to the mercury, many of which will form amalgams. The polarity of the mercury-drop electrode is now changed, so that it becomes the anode, and the applied potential is increased. As this reaches the oxidation potential of the dissolved metal, ions will pass into solution and so an increase in current will be observed. This will decline and stabilize, until the oxidation potential of other components is reached. The resulting voltammogram will then be a series of peaks at voltages corresponding to the individual components and the size of the peaks will be a measure of the amounts of those components present in the mercury drop and therefore in the original solution.

13.3.2 *Amperometric titrations*

The polarographic process can be used as a means for the detection of end points in titrations. A constant voltage corresponding to the diffusion-current region of the polarographic wave is applied between the micro-electrode (polarized) and the working electrode. As the concentration of the electro-active component is reduced during the titration, so the rate of diffusion to the microelectrode and hence current will be reduced. Consider the precipitation titration of lead ions with sulphate. As the titration proceeds, the amount of lead in solution, and hence the diffusion-controlled current, will fall until the end point is reached, when the current will remain constant as the excess sulphate is not electroactive under these conditions.

The technique can be extended using a pair of polarized electrodes with a solute which can be reversibly oxidized and reduced, where the amount of reduction taking place at the cathode equals that of oxidation at the anode. An

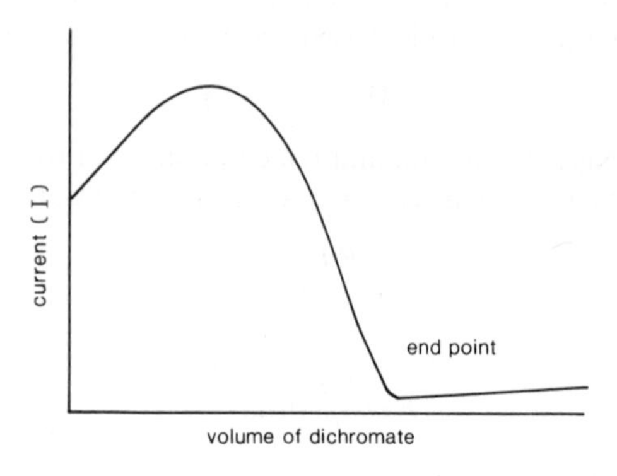

Figure 13.12 Amperometric ('Dead-stop') titration of Fe^{2+} ions with dichromate.

example of this technique is the oxidation of ferrous ions by dichromate (Figure 13.12). At the beginning, the current is low as there is only a very small concentration of ferric ions in solution and at the end point the same effect is observed as the ferrous-ion concentration is now minimal. The maximum current is reached near the half end-point titre where the concentrations of Fe^{II} and Fe^{III} are nearly equal. After the end point, no increase in current is observed as dichromate ions do not form a reversible couple under these conditions. Such titrations are often known as dead-stop titrations. It is also possible to have a situation where only the reagent forms a reversible system and then the converse would be observed, that is, with a constant current until the end point is reached and then an increase as the titrant is present in excess.

13.4 Potentiometric measurements

Potentiometric techniques involve measurement of the potential of an electrode, whose electrode reaction involves the analyte or titrant, so that the observed electrode potential will alter as the analyte changes. Remember that electrode potentials cannot be determined in isolation and so a correction must be made for the associated reference electrode (see Section 13.1.2).

13.4.1 *pH measurement*

The single most important application of potentiometric determinations is undoubtedly that of pH. The pH of a solution is simply defined as

$$\mathrm{pH} = -\log_{10} a_{\mathrm{H}^+}$$

so that any electrode whose potential responds to changes in hydrogen-ion activity of the solution with which it is in contact could in theory be used to determine pH. However, for practical reasons the range of such electrodes employed is greatly reduced.

13.4.1.1 *Hydrogen electrode.* The simplest electrode responsive to changes in a_{H^+} is the hydrogen electrode, based on the electrode reaction:

$$\mathrm{H}^+ + e^- \rightarrow \tfrac{1}{2}\mathrm{H}_2$$

Applying the Nernst equation and remembering that the standard electrode potential for the hydrogen electrode is zero by definition (see Section 13.1.2):

$$E = \frac{RT}{F} \ln \frac{a_{\mathrm{H}^+}}{(a_{\mathrm{H}_2})^{1/2}}$$

or

$$E = \frac{RT}{F} \ln a_{\mathrm{H}^+}$$

if the partial pressure of hydrogen gas, and hence approximately its activity, is fixed at one. The electrode is however, cumbersome to use and the electrode surface may be poisoned, affecting results, and is now no longer widely used.

13.4.1.2 *Quinhydrone electrode.* The quinhydrone electrode is based on the electrode reaction:

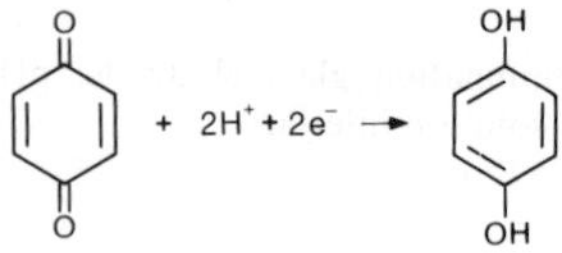

O

$$+ \ 2H^{+} + 2e^{-} \rightarrow$$

O

OH

OH

An equal concentration of quinone and hydroquinone in equilibrium is maintained by addition of a sparingly soluble equimolar complex of these components (quinhydrone). The Nernst equation then simplifies to

$$E = E^0 + \frac{RT}{F} \ln a_{H^+}$$

The electrode (platinum electrode placed in solution to be measured with added quinhydrone) is reliable over the pH range 1–8, but above this pH atmospheric oxidation of hydroquinone can take place, disrupting the equilibrium. In addition, quinone can react with nucleophilic compounds such as amines. Again this electrode is no longer widely used.

13.4.1.3 *Glass electrode.* The glass electrode is the most important pH-sensitive electrode and has now almost completely replaced other techniques. Its mode of operation is, however, somewhat different from these other electrodes. The glass electrode makes use of the property that when a membrane of certain glasses is subject to solutions of different a_{H^+} on each side, a potential difference develops across the membrane. This is thought to be due to an ion-exchange process involving exchange between sodium and hydrogen ions:

$$H^+_{soln} + Na^+_{glass} \rightleftharpoons H^+_{glass} + Na^+_{soln}$$

Thus, if the a_{H^+} is kept constant on one side of the membrane, usually inside a bulb, then the potential across the membrane will depend solely on the a_{H^+} of the external, test, solution. It is therefore simply a matter of measuring this potential, which is achieved by placing a reference electrode inside the bulb (Ag/AgCl) which will be at constant potential and a second reference electrode (calomel) connected via a salt bridge to the test solution. The only point in the system (Figure 13.13) where the potential changes is at the membrane/test-solution interface, which in turn depends on a_{H^+} of the test solution. The glass electrode, with its associated internal and external reference electrodes, is usually constructed as a single unit (combination electrode) which greatly facilitates its use.

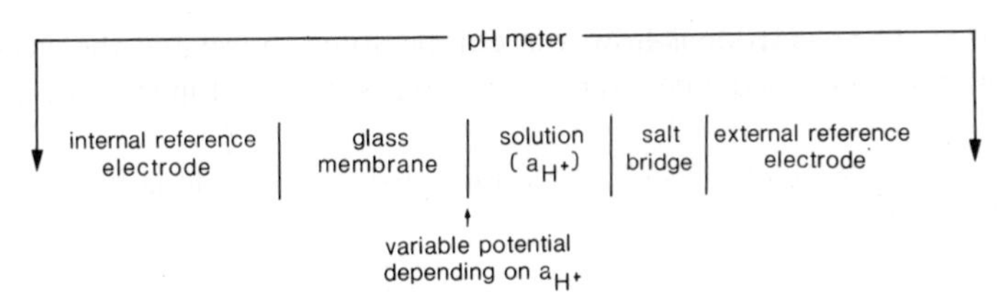

Figure 13.13 Components of combination glass electrode. pH meter responds to variable potential at glass membrane/test solution interface.

The glass electrode provides a rapid and reliable method of determining pH values over a wide range, but it does have a number of characteristics which need to be considered.

(i) Owing to the very high resistance of the glass membrane, sensitive recording equipment is required in which the very small currents that arise are amplified prior to estimation. This clearly increases the expense of the equipment.
(ii) The two surfaces of the membrane are rarely identical, so that, even when solutions of identical a_{H^+} are placed on either side of the membrane, a small potential difference is observed. Furthermore, this residual value (asymmetry potential) changes with age and condition of the membrane surface. Thus, theoretically calculated potentials cannot be used to calibrate the pH meter and this must be carried out with buffers at frequent time intervals and also ideally at different pH's.
(iii) As the process at the membrane surface is one of ion exchange, it is not surprising to find that this is not completely specific. Thus, at low a_{H^+}, that is, high pH, it is found that ions such as Na^+ in solution will compete with H^+ for interaction with the surface and so lower values are recorded. This effect (alkaline error) may be greatly reduced by using specialized membranes, for example those in which the majority of the Na^+ ions have been replaced with Li^+.
(iv) The glass electrode is also inaccurate at very low pH's (below 1), where the water activity in the hydrated surface layer of the membrane is thought to be affected.

The importance of accurate measurement of pH in numerous areas within the biological sciences must be self-evident, for example in controlling enzymic reactions, preparation of physiological fluids, protein precipitation, ion-exchange chromatography and electrophoresis. It is also a prerequisite for pH-stats in which the pH is kept constant by automated addition of acid or alkali, for example to counteract the effects of metabolites from growing organisms. Another important application is in the end-point determination in acid–base titrations and it makes possible the use of automatic titrators which are able to titrate to a predetermined end point (that is, pH).

13.4.2 *Ion-selective electrodes*

The observation that glass electrodes are not completely specific to H^+ ions (alkaline error) led to the development of other electrode systems which are more selective towards other ions. There are several types available.

13.4.2.1 *Glass electrodes.* Here the composition of the glass has simply been altered to increase the selectivity to particular cations. The alkaline error of a pH electrode can be reduced by introducing lithium into the glass and in an analogous manner the selectivity towards alkali metal cations can be increased by introducing aluminium. A typical glass composition for a sodium electrode might be: Na_2O, 11%; Al_2O_3, 18%; SiO_2, 71%. Electrodes of this type are also available for Li^+, K^+ and Ag^+. There will always be a certain amount of interference between similar ions and this should be established for each particular analysis.

13.4.2.2 *Solid-membrane electrodes.* In these cases, the glass membrane is replaced by a dispersion of a salt containing the ion of interest in a suitable silicone rubber matrix. In the original Pungor (1965) electrode, silver iodide was dispersed in a silicone polymer matrix to form an iodide responsive electrode, the activity of iodide ions, and hence surface charge, being kept constant on the inside of the membrane by a solution of 0·1 M potassium iodide. The range of these heterogeneous membranes has now been extended to include many other anions, such as Cl^-, Br^-, CN^- and S^{2-}. However, the technique is not restricted to anion electrodes and electrodes for Ag^+, Pb^{2+}, Cd^{2+} and Cu^{2+} have been produced. The membrane may be replaced by a single crystal of the appropriate silver halide, or at least a casting of the salt, to form electrodes for the halides Cl^-, Br^- and I^-. An electrode for F^- ions has also been developed, based on a single crystal of lanthanum fluoride.

13.4.2.3 *Liquid-membrane electrodes.* Here a liquid ion exchanger in the form of the ion of interest is dissolved in an organic solvent and separated from the test solution by a porous diaphragm; an example would be the calcium salt of didecyl hydrogen phosphate, dissolved in di-*n*-octylphenylphosphonate to form a calcium responsive electrode (Ross 1967). A novel electrode of this type is the K^+ electrode in which valinomycin (an antibiotic) is used as the ion-exchange material. This electrode is particularly useful as it shows a very high degree of selectivity over Na^+ and H^+ ions.

13.4.2.4 *Mode of use of ion-selective electrodes.* The electrodes of all three types are used in an analogous manner to pH responsive electrodes, often as combination electrodes with built-in reference electrodes and integral salt bridges (see Figure 13.13). The electrode response is governed by the Nernst equation:

$$E = K + \frac{RT}{nF} \ln a_{M^{n+}}$$

where $a_{M^{n+}}$ is the activity of the cation of interest. For a monovalent cation (such as Na^+) this may be reduced to:

$$E = K - 0{\cdot}059\,\text{pM}$$

where $\text{pM} = -\log a_{M^{n+}}$, in an analogous manner to the pH scale. In practice, these electrode systems are never specific, hence the use of the more correct phrase ion-selective electrodes, and so there will also be a further contribution, hopefully small, from other ions in the sample. An idea of the magnitude of such interferences should be obtainable from the electrode specification, but when the electrode is calibrated with standard solutions of the ions of interest, the effect of potential interfering ions should also be investigated. It should also be noted that the electrodes respond to the activities of the ions, rather than their concentrations. Therefore ion-selective electrodes will only respond to (hence measure) the proportion of the element that is present as free ions and will ignore, for example, protein-bound material. A further complication may arise if the ionic strength of the solutions used for calibration is markedly different from that in the sample, which is often the case with biological samples. One way around this problem is to use the technique of standard additions, in which the sample matrix is effectively used to make up the calibration standards, hence affecting the standards and analyte to the same extent (see Section 11.2.2).

In addition to the use of ion-selective electrodes for direct analysis for example, the determination of metals in diluted plasma, or nitrate/nitrite in river water, they may also be used to monitor titrations. A case in point would be the use of an Ag^+ electrode to detect the end point in an $AgNO_3$ titration for chloride ions; as the end-point is reached, the Ag^+ concentration and hence the response from the electrode will increase; before this point any added Ag^+ will be removed from solution by precipitation of AgCl.

13.4.3 *Potentiometric titrations*

The use of potential measurements to determine the end point in acid–base titrations has already been mentioned (Section 13.4.1), where the glass electrode is used to follow pH. There are, however, a number of other volumetric techniques where end points can be determined potentiometrically, namely precipitation titrations and redox titrations.

13.4.3.1 *Precipitation titrations.* The end point in a silver nitrate titration of chloride ions can be determined by conductivity measurements (Section 13.2.2) or by the use of indicators such as dichromate (forms red/brown silver chromate in presence of excess Ag^+). Additionally, the end point may be

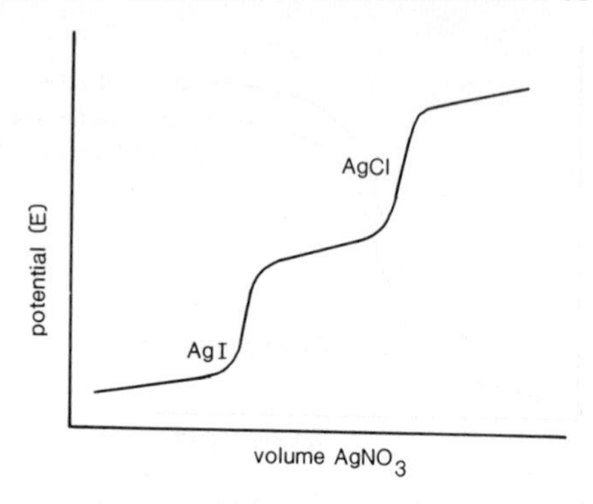

Figure 13.14 Precipitation titration followed potentiometrically using a silver electrode.

determined with a silver electrode, coupled to a suitable reference electrode. Consider the titration of a sodium chloride solution, possibly an aqueous food extract or diluted plasma, with silver nitrate. As Ag^+ ions are added, they will react with Cl^- to form insoluble AgCl so that in the early parts of the titration the Ag^+ concentration in solution will remain very low and reasonably constant and so therefore will the potential of the silver electrode. As the end point is passed, there will be a sudden increase in Ag^+ concentration and this will be reflected by a rapid change in electrode potential, allowing the end point to be determined. The technique can be extended to solutions of mixed halides, where separate end-points corresponding to each anion will be observed, provided there is a sufficient difference in the solubility of the silver salts formed. Thus a mixture of iodide and chloride could be titrated with the result shown in Figure 13.14. The iodide is titrated first, as silver iodide is less soluble than silver chloride. The situation would not be so clear if, for example, chloride and bromide were present, as these have more similar solubilities.

13.4.3.2 *Redox titrations.* Here again, redox titrations can often be carried out using indicators to determine end points, either with an external indicator (a dye which changes colour as the redox potential of the solution changes) or where the reagent changes colour itself (for example, permanganate changing from purple to colourless in acid solution as it is reduced). However, in many instances, potentiometric methods are to be preferred as they allow automated titrations to a predetermined potential. Consider the titration of Fe^{2+} ions with Ce^{4+}. The following reactions will take place:

$$Fe^{2+} \rightarrow Fe^{3+} + e^-$$

$$Ce^{4+} + e^- \rightarrow Ce^{3+}$$

In the first part of the titration, that is, before the end point is reached, Fe^{2+} and Fe^{3+} ions will be in solution and their relative amounts, which will alter as the titration proceeds, will dictate the potential of the solution. Near the end

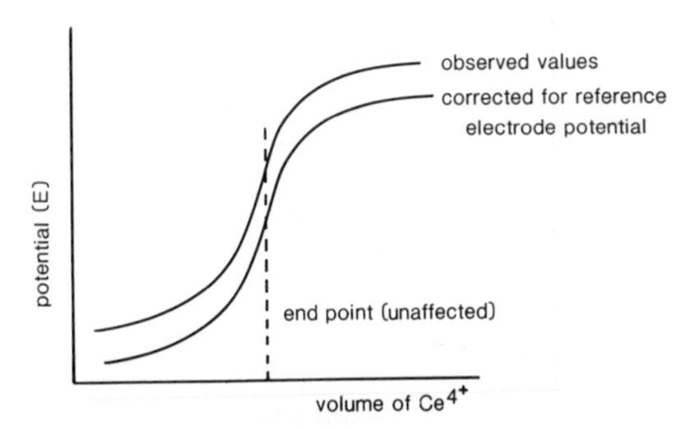

Figure 13.15 Potentiometric titration of Fe^{2+} ions with Ce^{4+} ions, showing no change in end-point on correction for reference electrode potential.

point, the ratio Fe^{3+}/Fe^{2+} will alter very rapidly and so therefore will the potential. After the end point there will be no Fe^{2+} in solution and the potential will be governed by the ratio of Ce^{4+}/Ce^{3+} as excess Ce^{4+} is added. A potential curve, as shown in Figure 13.15, is obtained where the end point is the point of inflexion and this can be shown to occur at the mean of the electrode potentials of the two redox couples involved.

$$E_{\text{end point}} = \tfrac{1}{2}(E^0_{Fe^{3+}/Fe^{2+}} + E^0_{Ce^{4+}/Ce^{3+}})$$

This allows the end point to be theoretically determined and these values can be used to allow titrations to be stopped at a particular potential, rather than needing the entire curve to find the inflexion point. Where the redox potentials are not known under the conditions of the titration, the end point may be more precisely determined by derivative techniques.

In potentiometric titrations, if the shape of the curve alone is being used to determine the end point, it is not necessary to correct the observed potentials for the reference electrode potential, as this will only shift the curve in the *y*-direction (Figure 13.15) and will not influence the end-point value.

13.4.4 *Oxygen electrodes*

In the galvanic type of electrode, the current output is directly proportional to the oxygen partial pressure. The electrode consists of a glass tube which supports the silver cathode and lead anode. A polypropylene (or similar polymer) is sealed to the base of the tube, which is filled with a mixed electrolyte of sodium and lead acetates in aqueous acetic acid (Borkowski and Johnson 1967). The electrode assembly is placed in the solution of interest and oxygen diffuses across the membrane where the following reactions take place:

$$\text{Reduction: } \tfrac{1}{2}O_2 + H_2O + 2e^- \rightarrow 2OH^-$$

$$\text{Oxidation: } Pb \rightarrow Pb^{2+} + 2e^-$$

$$\text{Overall: } \tfrac{1}{2}O_2 + Pb + H_2O \rightarrow Pb(OH)_2$$

The resulting current is then dependent on the partial pressure of oxygen in the electrolyte and hence concentration within the sample. Lead is consumed during the process and so the electrode does have a finite life, usually about 12 months. The electrode must be calibrated, in an analogous manner to a pH electrode, and this is often carried out at 0% and 100% oxygen saturation.

Alternative cells are available for determining oxygen based on electrolytic reduction (Clark electrode) and these also make use of a gas-permeable membrane.

Dissolved oxygen is a very important parameter in many biochemical areas. For example, the amount of oxygen dissolved in a fermentation broth may be critical for the growth of particular organisms. If the level of oxygen can be continuously monitored, any deviation from the required value can be automatically corrected, for example by pumping air through the system.

Environmentally, the amount of dissolved oxygen in river waters is a good indication of their 'health'. As the level of pollution increases, the oxygen is used up by growing aerobic organisms, until the water is oxygen-depleted and anaerobic organisms will then start growing, leading to foul water. Pollution of water systems can also be judged by their biochemical oxygen demand (BOD) which can be measured from the difference in dissolved oxygen before and after a predetermined incubation. The higher the level of organic pollution, the more oxygen will be used up during the incubation. Routine measurements of BODs are greatly facilitated by relatively simple methods of determining dissolved-oxygen values, as for example with oxygen electrodes.

Apart from oxygen electrodes, similar gas-sensing probes are available for NH_3, CO_2, HCN, HF, H_2S, SO_2 and NO_2. The first two of these, as well as the oxygen electrode already mentioned, are particularly important for monitoring the progress of fermentations, either in batch or continuous fermentors.

References

Bassett, J., Denney, R.C., Jeffery, G.H. and Mendham, J. (1983) *Vogel's Textbook of Quantitative Inorganic Analysis*, 4th edn, Longman, London, ch. XVI.
Borkowski, J.D. and Johnson, M.J. (1967) *Biotech. and Bioeng.* **9**, 635.
Ives, D.J.G. and Janz, G.J. (1961) *Reference Electrodes*, Academic Press, New York.
Murthy, G.K. (1974) *CRC Critical Rev. Environm. Control* **5**, 1–37.
Pungor, E., Havas, J. and Toth, K. (1965) *Z. Chem.* **5**, 9.
Ross, J.W. (1967) *Science* **156**, 1378.
Ramette, R.W. (1981) *Chemical Equilibrium and Analysis*, Addison-Wesley, Reading, MA.

Further reading

Frieser, H. (1978) *Ion-Selective Electrodes in Analytical Chemistry*, vol. I, Plenum, New York
Heyrovsky, J. and Kuta, J. (1965) *Principles of Polarography*, Academic Press, New York.
Rossotti, H. (1969) *Chemical Applications of Potentiometry*, Van Nostrand Reinhold, New York.
Weissberger, A. and Rossiter, B.W. (1971) *Physical Methods of Chemistry Parts IIa and IIb: Electrochemical Methods*, Wiley-Interscience, New York.
Westcott, C.C. (1978) *pH Measurements*, Academic Press, New York.

Index

Abbreviations used:

AAS	atomic absorption spectrometry
AFS	atomic fluorescence spectrometry
ESR	electron spin resonance spectroscopy
PES	flame emission spectrometry
GC	gas chromatography
GLC	gas—liquid chromatography
HPLC	high performance liquid chromatography
IR	infrared
LC	liquid chromatography
MS	mass spectrometry
NAA	neutron activation analysis
NMR	nuclear magnetic resonance
TLC	thin layer chromatography
UV	ultraviolet